China’s Belt and Road Initiatives

Wei Liu
Editor

China's Belt and Road Initiatives

Economic Geography Reformation

With Contributed by Jianxiong Ge, Angang Hu, Yifu Lin, Liang Qiao

上海交通大學出版社
SHANGHAI JIAO TONG UNIVERSITY PRESS

Springer

Editor
Wei Liu
Guangming Daily
Dongcheng, Beijing
China

Translated by Jun He and Jia Lao

ISBN 978-981-13-0100-1 ISBN 978-981-13-0101-8 (eBook)
https://doi.org/10.1007/978-981-13-0101-8

Jointly published with Shanghai Jiao Tong University Press, Shanghai, China

The print edition is not for sale in China Mainland. Customers from China Mainland please order the print book from: Shanghai Jiao Tong University Press.

Library of Congress Control Number: 2018939139

B&R Book Program

Printed on acid-free paper

This Springer imprint is published by the registered company Springer Nature Singapore Pte Ltd.
part of Springer Nature
The registered company address is: 152 Beach Road, #21-01/04 Gateway East, Singapore 189721, Singapore

Preface

Last July, coordinated by Guangming Daily's Shanghai bureau, Guangming Daily's Forum Edition (hereinafter referred to as the Forum Edition) and Shanghai Jiao Tong University Press co-hosted the Guangming Yuanji Forum. Invited to speak at the first lecture were Mr. Ji Weidong, Dean of SJTU's Law School and Mr. Cheng Zhaoqi, Director of SJTU's Tokyo Trial Research Center. Both spoke on the trial of the century under the title *The Tokyo Trial is a Civilized One*. After publication of the Forum Edition, the articles were then reprinted by many websites and were well received.

Earlier this year, while in Shanghai for business, I revisited the SJTU Press and consulted with them on the future development of Guangming Yuanji Forum and profound cooperation on publication. Soon afterwards, Ms. Liu Peiying, the standing vice editor-in-chief telephoned Ms. Liu Qian, editor-in-chief of the Forum Edition, to say that they had read the scholars' speeches on the Belt and Road on the Guangming Forum and appreciated the articles, so the SJTU Press suggested that they should compile these speeches into a book and publish it. After much deliberation, we gladly agreed.

The SJTU Press (to be referred to as the Press) is always good at approaching critical subjects of social life, science and culture. For example, *The International Military Tribunal for the Far East: A Trial Record* and the 50 volumes of *Evidence and Documents from the International Military Tribunal for the Far East*, both are national projects assigned to the Press. Last year, they designed and published *Xi Jinping: Wit and Vision*. This brilliant book then became a best-seller and was well received by the public. And this time, visionary as they are, the Press has chosen the Belt and Road series on the Forum Edition.

The Belt and Road initiative is a strategic conception put forth by President Xi Jinping. And it bears realistic significance that is vital to the great revival of the Chinese nation. This conception was well received as soon as it is proposed. Professor Liang Qiao of National Defence University of People's Liberation Army described it as: "By far the best strategy that matches China's superpower status." And Prof. Angang Hu of Tsinghua University wrote: "Economically speaking, the Belt and Road initiative is a revolution of gigantic scale that will recreate economic geography of China and the world. From a diplomatic point of view, it will give rise to an era of win-win cooperation". When interviewing Prof. Angang Hu, a Korean reporter pointed out that some saw the proposal as the Chinese version of the Marshall Plan, to which the professor replied: "The Belt and Road initiative can be called the China Plan or the Xi Jinping Plan but it is definitely not the Marshall Plan, for the latter only benefited developed countries whereas the Belt and the Road mainly focuses on the developing world with the developed included as well. It is open to all nations with a spirit of mutual benefit and win-win".

Guangming Daily is a national newspaper run and managed by the CCCPC. Since its establishment in June 1949, the newspaper has been operating on the principle that we are a newspaper whose readers are mainly intellectuals, we report news of the intellectual world and print stories of experts and scholars. The Forum, established in 2006, is one of our star columns. It is a one-page column published once a week, running exclusively speeches and academic lectures of experts and scholars. The column actually benefits from *Guangming Daily*'s good reputation because every time an editor asks a prominent scholar to give a speech on a certain subject they all say yes with delight. Or when an editor comes across information about a scholar's speech and wants to print it on the newspaper they'd usually get the approval with just a phone call or an email. The more prominent a scholar is, the more serious his or her attitude is towards academics. I once saw this speech by Tang Min and Yifu Lin on the Belt and Road and they had some really fresh ideas. Tang Min indicates that there are huge differences in economy, politics, law and society among countries along the Belt and Road regions and that means great risks. There is also the possibility of strategic conflicts between superpowers caused by challenging current international order. The handling of said issues is of vital importance. Yifu Lin, on the other hand, thinks the Belt and Road initiative offers a new model of development and cooperation based on construction of infrastructure and industrial transfer therefore has considerable significance to global development. Tan Min, whom I've had the privilege studying within the Tsinghua Harvard class, is a well-known economist and a State Department Counselor. So, I called Tang Min and asked him for a contribution. He sent me the revised draft the next morning. When I asked Yifu Lin before and Angang Hu this time they sent me emails en route, discussing in a serious manner about revising a certain paragraph, sometimes even for a single word.

Zhang Yuanji was one of the founders of the publishing industry in modern China, the first Dean of the Translation Academy of Nanyang College, presently known as the Shanghai Jiao Tong University Press. He was also the Founder of the Commercial Press and the second President of SJTU. On 15 December 2007, *Guangming Daily* published an article titled *Zhang Yuanji: Life and Struggles of a Publisher*, which was a long report on this intellectual giant. Of course, the newspaper has done coverage of Zhang Yuanji before and there's more to come since.

In fact, there is a historical origin for *Guangming Daily* and SJTU's co-hosting the Guangming Yuanji Forum.

The Japanese invasion of China came with not only ruthless military strikes but also an elimination of Chinese culture. In 1932, after Japanese planes bombed the Commercial Press headquarters and Oriental Library, a total of some 460,000 books, including countless rare or unique copies, were burnt to ashes. This war crime against China on Japan's part infuriated millions and broke their hearts. So, eighty years later, when the Guangming Yuanji Forum opens with a lecture on the Tokyo Trial with successors of Zhang Yuanji testifying against Japanese invaders it can be viewed as a memorial of Mr. Zhang.

As early as the founding of the People's Republic of China, leaders of New China had invited Zhang Yuanji to Beijing for the first-ever Chinese People's Political Consultative Conference. On September 24, the eve of the conference, Xie Gongwang, a correspondent with *Guangming Daily* paid a special visit to Zhang Yuanji, asking about his personal story as well as his opinion on the new government. It was then that Zhang Yuanji gave Xie Gongwang three copies respectively of *Plain Words*, *New Proverbs on Family Management* and *Songs of Extraordinary Women*, while telling him that, if the interview was to be printed, the draft must be presented to him for approval. This direct contact with *Guangming Daily* was recorded in Zhang Yuanji's personal journal.

This co-publication by *Guangming Daily* and SJTU is a demonstration of perspectives of a number of Chinese intellectuals. I believe the book will generate a benign social impact.

Not only will the Belt and Road initiative introduce to the world the traditional Chinese idea of harmony, but also the economic concept of mutual benefit and win-win, which also has its roots in Chinese traditions.

How will the Belt and Road initiative carry out? How will hundreds of countries in the region respond? What influence will it have on global politics and economy? You, dear reader, will know what scholars and experts think when you open this book.

Dongcheng, Beijing, China
August 2015

Wei Liu
Vice Editor-in-Chief
Guangming Daily

Contents

The Silk Road: Historical Geographic Background and Outlooks 1
Jianxiong Ge

The Belt and Road: Revolution of Economic Geography and the Era of Win-Winism 15
Angang Hu

The Belt and Road and a New Model of International Development Cooperation 33
Yifu Lin

Grand Strategic Thoughts with a Global Vision 39
Liang Qiao

The Belt and Road as the Lead to a New Era of Opening up 49
Min Tang

Westward Opening: A Perspective of the Silk Road Economic Belt 57
Wenbin Xia

How the Belt and Road Initiative Expands the Diplomatic Arena for China 67
Peisheng Yao

The Idea of "Transparent Ocean" Drives Future Expansion for China 77
Lixin Wu

Guided by Shipping and Guarded by Law—Legal Thoughts on Building the Modern Maritime Sild Road 89
Wei Zhao

Let the East Gale Unfurl the Lonely Sail—Historical Thoughts on the Modern Maritime Silk Road 101
Wang Jie

Diversity of Chinese Culture and the Belt and Road Initiative 115
Shiyuan Hao

Position and Role of Yunnan in China's Westward Opening-up Strategy 129
Jianhua Fan

On Further Promoting Cooperation and Development of Education Among Countries Along the Belt and Road 141
Zhenyuan Qu

Appendix A: Vision and Actions on Jointly Building Silk Road Economic Belt and 21st-Century Maritime Silk Road 157

Appendix B: Promote Friendship Between Our People and Work Together to Build a Bright Future—Speech at Nazarbayev University 171

Appendix C: Join Hands to Build the China-ASEAN Community of Common Destiny—Speech at the Indonesian Parliament 177

Contributors

Jianhua Fan President of Yunnan Federation of Social Science Circles.

Jianxiong Ge Member of the 12th CPPCC Standing Committee, Professor of Fudan University.

Shiyuan Hao Assistant to the President, Chinese Academy of Social Sciences.

Angang Hu Renowned Economist, Dean of the National Research Institute of Tsinghua University.

Wang Jie Professor and doctoral supervisor at the Transportation Management College of Dalian Maritime University.

Yifu Lin Renowned Economist, Former Chief Economist and Senior Vice President of the World Bank.

Liang Qiao Professor at National Defence University, Air Vice Marshal, China Air Force.

Zhenyuan Qu President of China Association of Higher Education.

Min Tang Deputy Secretary General of China Development Research Foundation, Counselor of the State Council.

Lixin Wu Professor at Ocean University of China, Academicia at the Chinese Academy of Sciences.

Wenbin Xia Deputy Director of CCP Committee of Shihezi University, Director of the Collaborative Innovation Center of Central Asian Civilization and Westward Opening.

Peisheng Yao Former China's ambassador successively to Kyrgyzstan, Latvia, Kazakhstan and Ukraine.

Wei Zhao Professor and doctoral supervisor at the Law School of Dalian Maritime University.

Summary

The construction of the Silk Road Economic Belt and the Twenty-first Century Maritime Silk Road (hereinafter referred to as the Belt and Road) is a significant strategic conception put forth by the Central Committee of the Communist Party of China, led by General Secretary Xi Jinping, in order to further improve China's economic openness. This book was compiled from a series of lectures on the Belt and Road, originally released on the *Guangming Daily*'s "Guangming Forum" (hereinafter referred to as the Forum). A total of thirteen well-known scholars and experts, including Jianxiong Ge, Angang Hu, Yifu Lin and Liang Qiao, co-wrote this book, which offers a comprehensive interpretation of the Belt and Road strategy and its influence from perspectives of politics, history, economy, military, diplomacy, culture, technology and law.

The Silk Road: Historical Geographic Background and Outlooks

Jianxiong Ge

The Origin of the Silk Road Concept

Speaking of the Silk Road, we should first understand its historical facts, and why such a road existed in the first place, which was closely related to the historical conditions at that time. Marxism historical materialism has a basic idea, that all kinds of cultures or thoughts, including religion, politics or scientific activities, are all based on the daily lives of the contemporary people, that is, the material conditions. Especially at the time of low productivity and technologies, all activities were subject to serious geographical constraints. Therefore, we must understand the geographical environment at that time, not only the natural one, but also the cultural one.

It is generally understood that the Silk Road started from Luoyang and Chang'an of ancient China, went all the way through the Hexi Corridor to its terminal of Dunhuang before trifurcating: the north route to the Tianshan Mountains and Urumqi; the middle route from Turpan to the west, through the southern edge of Xinjiang, Hotan to Central Asia, then through Central Asia to Persia (Iran) and finally to Europe; and the south route along the Pamir plateau today Into Pakistan and India. Some scholars in China transplanted the Silk Road concept onto other routes, such as the "Southwest Silk Road", "Southern Silk Road", "Northern Silk Road", "Grassland Silk Road", etc., without being widely accepted. Now the question is, why are these routes put on the name Silk Road? How did such a name come about? Many people thought this concept comes from ancient China, and many got it confused. The Silk Road we talk about now is not just any road

J. Ge (✉)
Fudan University, Shanghai, China

W. Liu (ed.), *China's Belt and Road Initiatives*,
https://doi.org/10.1007/978-981-13-0101-8_1

or traffic route, but a specific concept. Who coined this term? Ferdinand von Richthofen, a German geographer. He studied the geography Central Asia and the western part of China in the 1860s, for about three years, and studied historical records in China. After going back to Germany, he published his research, where he proposed that in the 2nd century AD, there was a business route from Luoyang and Chang'an to Samarkand in Central Asia, and the main business was silk, sent from China to Central Asia, West Asia, and finally Europe. Since silk was the main business and of the greatest impact, so he named the route Silk Road. Therefore, you won't see Silk Road mentioned anywhere in *The Twenty-Four Histories, Siku Quanshu* or any book about China before 1870.

The Silk Road Was Paved by People of Multiple Nations

Who is this way to open it? In March 2015, the National Development and Reform Commission, Ministry of Foreign Affairs, and Ministry of Commerce of the People's Republic of China described it well in its vision, "More than two millennia ago, the diligent and courageous people of Eurasia explored and opened up several routes of trade and cultural exchanges that linked the major civilizations of Asia, Europe and Africa, collectively called the Silk Road by later generations. It contributed greatly to the prosperity and development of the countries along the Silk Road. Symbolizing communication and cooperation between the East and the West, the Silk Road Spirit is a historic and cultural heritage shared by all countries around the world." Numerous evidence has proven that the road exists since long ago. In the Tomb of Fu Hao of the Shang Dynasty found a number of jade artifacts, which have been identified as Hetian jade and Kunlun jade. The fact that they are found in tombs of the Shang Dynasty shows that such a road existed at that time (17th century BC to 11th century BC). Somebody shipped the jade to today's Henan Province three thousand years ago—how would that be possible without a road? There are even earlier evidences that this road has long existed. Taking a different perspective, present-day Xinjiang was influenced not only by the Central Plain, but also by the West. Ancient corpses found in Xinjiang was obvious Caucasian, and some of those were certainly from Europe. As Mr. Ji Xianlin once said, Central Asia was at the crossroad of cultural exchange of the East and the West. There must have been a road—whether it was natural or paved by the ancient—the road must have existed.

"2nd century AD", as mentioned by Richthofen, was the time that Zhang Qian of the Han Dynasty went on a mission to the Western Regions. Yet his journey was not to exchange goods, or to export Chinese silk, or to do foreign trade, but for political and military purposes. Before his journey, the Huns was a huge threat to the Han Dynasty. When the Western Han Dynasty was founded, the Huns can even threaten Chang'an, its capital. As its strength recovered, especially at the time or Emperor Wu, the Han Dynasty finally had the ability to launch a comprehensive counterattack against the Huns. However, facing the cavalry-based Huns with great mobility and agility, the Han Dynasty struggled to defend the long border from Liaodong Peninsula to the west bank of Yellow River. So Emperor Wu wanted to completely destroy the Huns' effectives. He received information that Yuezhi, one of the two peoples originally living in Qilian Mountains, under the oppression of the Huns, was forced to move to today's Afghanistan, known as Daxia. Knowing the running feud between Daxia and the Huns, Emperor Wu wanted to set up an alliance with Daxia so that they would attack the Huns from the west and the Han Dynasty attack from the east. Emperor Wu recruited messengers to go to the Western Regions on behalf of the court, and Zhang Qian applied. Zhang Qian was captured by the Huns just stepping out of Han's control, and detained for 10 years. But he never forgot his mission. Getting a chance to escape, he did not flee back to his hometown, the Han Dynasty, but continued his mission and finally arrived at Daxia. Yet neither his political nor his military mission was accomplished. The king of Daxia was enjoying himself with bigger territory and stronger nation than he had at Qilian mountain, and wasn't willing to start a war with the Huns. *Records of the Grand Historian* described the mission of Zhang Qian "no point" which later became the Chinese idiom "不得要领". With "no point" staying there, Zhang Qian headed back, and was detained again by the Huns on his way. This time he quickly escaped and returned to the Han Dynasty. Although he did not complete Emperor Wu's mission, his briefing and experience intrigued great interest of the Emperor. At that time, the military situation had undergone drastic changes. Han Dynasty had defeated the Huns, the door to the Western Regions had been opened and the Hexi Corridor was under the control of the Han Dynasty. It was possible to travel from Chang'an through the Hexi Corridor and go directly to the Western Regions. So, Zhang Qian took a second mission. This time he had a mighty troop of hundreds of people, as he knew it would be impossible to go to those countries all by himself. He sent his deputies to various countries with numerous gifts. In his era, Emperor Wu deemed himself as the almighty ruler of the world, so he wanted to attract other people to the Central Plains. Among the gifts there was silk, which was relatively portable. Exotic to the Western Regions, silk was exclusively produced in the Han Dynasty. Yet the silk was not for trading, but as a reward. As Zhang Qian and his deputies travelled, a large number of Central Asian countries and local people were aware of the beauty and preciousness of silk, which then expanded silk trading to Central Asia and West Asia through merchants. That is why Richthofen called it the Silk Road.

A statue of Zhang Qian as the envoy to Central Asia

Since then, the Emperor of Han sent more messengers to the Western Regions, and some of them misappropriated those gifts—a free gift from Han's treasury became their own commodity for profit, which, from an objective perspective, expanded the trade. Yet those actions would not amount for a Silk Road. The main drive was the strong demand for this rarity from Central Asia, Persia and the Roman Empire. Good silk products were exclusively from Han, and rich Romans had nowhere else to get it. Silk clothing and silk ornaments were symbols of extraordinary wealth. The Romans had a tradition for business, so the strong demand matched up with the purchasing power, giving a strong drive to this route of trade. Therefore, it was not built or intentionally driven by the Han Dynasty—it was not China that actively exported silks. People along the Silk Road also lacked commercial exchange.

External Factors to Consider When Building the Silk Road

If we want to build the Belt and Road today, some external factors must be considered. For example, do Kazakhstan and Kyrgyzstan have the same enthusiasm, the same desire and the same interest as China do? A historical Silk Road doesn't mean we can build such a route today or it would be possible to build a Silk Road economic belt. That would be a misreading of the history. Our situation today is completely different. China has always considered itself to be "the center of world" and there is no demand to anybody else. But that's not the case today. Historically China was not actively making use of the Silk Road, or gaining major benefits from maritime trading. If we are to build the Belt and Road today, we cannot make such ineffective moves. We have to consider how to invite positive response and sustained motivations.

For example, a few years ago we signed an agreement with Kazakhstan to establish a Sino-Kazakhstan Trade City. China finished our side, yet there was literally no action from Kazakhstan. Why? They had their own considerations. Kyrgyzstan faced a similar situation. From the Kashi to the border, the elevation goes up to 3400 m—it snows in autumn time. Going past the border, there is a mountainous depopulated area of more than 100 km. How can they trade with China? I went there in 2003, and we saw trailer trucks, all full of home appliances, or even furniture—those countries would buy whatever China sent, but the trucks all came back empty. That was not a trade. Even if Kazakhstan is rich in cotton, but they can sell to whoever offer the highest price, not necessarily to China. They have oil and gas, but those are transported through pipelines. There are minerals, but not for border trading. Some measures have to be taken to motivate trading. In the Boao Forum, President Xi Jinping empathized mutually beneficial win-win situation and the establishment of a community of shared interests. How to create the demand from other countries? That's not easy. Take Pakistan. President Xi went there and we signed many contracts, such as nuclear power plants, which was reported to be over USD15 billion. But if their own economy stays weak for a long time, can this investment on infrastructure sustain? The economic belt cannot be a unilateral thing, but has to be bilateral or multilateral. Simply quoting the Silk Road in the history does not help us building it.

Often Broken, the Silk Road Depends on the Strength of China

The Silk Road is not always unimpeded, or even more often blocked. It only works when the regime on Central Plains firmly controls Central Asia, otherwise it would be more of a regional or local route rather than a thoroughfare between the East and the West. According to historical records, the Western Han Dynasty set up Protectorate of the Western Regions on 60 BC, covering an area of about 2 million square kilometers in present-day Xinjiang and Central Asia, which marked for the first time that Xinjiang belongs to China. But we also noted that, although the Western Han Dynasty had the sovereignty, it did not set up regular administrations such as counties or prefectures, because it was too far away from the Central Plains and too costly to administer. If the Emperor sent an official to today's Yanqi in Xinjiang (the usual headquarters of Protectorate of the Western Regions), it would be such a long trip that he might be killed on the road. There was no convenient measure of information transmissions, and it would probably be too late when the herald arrived. Therefore, the Western Han Dynasty can only be a guardian there. There were only a few thousand soldiers and a small group of officials to cover such a vast area, and they had to farm to feed themselves. How did they rule? By the authority of the central government and the courage of the officials.

With the special geographical conditions of Xinjiang, there was no way to implement the administrative hierarchy as in the Central Plains. At first there were 36 countries, and later 50–60 countries in the West Regions. Why so many? Because many places are completely isolated oasis. Seen from the plane, Xinjiang is like a big sandbox—full of the plateau, Gobi, desert, and the green or black blobs are the oasis—there is water where people can grow crops and settle. Big oasis can support tens of thousands of people, while small one can only accommodate hundreds. They would be bigger or smaller countries in the ancient times. For example, Wusun, the largest country in the Western Regions, had 300,000 people, while the smallest country had only 100 people. Those countries were far apart with depopulated areas in between. If dispute arose, the administration can only mediate, or command the forces around. When natural disasters hit, they were on their own and could not rely on the central government. Only such a loose administration was possible, which might work when things went well, yet unsustainable if the central government was weak or the officials were incompetent. At the end of the Western Han Dynasty, Wang Mang seized the throne and the Protectorate was dismissed. After the establishment of the Eastern Han Dynasty and the crackdown of internal separatist forces, another Protectorate of the Western Regions was set up, but with a different name (Chief Official of the Western Regions). In the 200 years of the Eastern Han Dynasty, the prefectural government of the Western Regions was set up and dismissed for three times, and only intermittent afterwards.

The route was working when the Western Regions was in the control of the Han Dynasty, and broken when it wasn't. The Silk Road, as defined by Richthofen, didn't last the entire time of the Western and Eastern Han Dynasty, but more specifically the time with stable regime. This is the same case after the Eastern Han Dynasty. For example, the Northern Wei Dynasty unified the north, then relocated its capital from Pingcheng to Luoyang. The Western Regions were under its control all the time, so the foreign trade was prosperous, and merchants from Daqin (the Roman Empire) can brought large amount of merchandise to China. But when there was war, not only Luoyang but the whole area along the route was in depression. The Tang Dynasty controlled Central Asia at its early establishment, and set up an administrative center—Suyab. It was an important stronghold of the Tang Dynasty in the Western Regions, as military headquarters and civil administration. With such large area under control, Tang Dynasty needed this artery for internal traffic. However, the good situation didn't last. After the An Lushan Rebellion broke out, the army moved eastwards to fight against An Lushan and Shi Siming. As the control on the Western Regions weakened, the Tibetans expanded from Tibet to Xinjiang and even seized the Hexi Corridor. Uighurs originally living in the Mongolian plateau moved westward to the Tarim River Basin, and became today's Uighurs after mixing with local people and absorbing others. Later the Tibetans withdrew from Xinjiang and Hexi Corridor, and the Arabian expansion spread Islam to Xinjiang. The Silk Road from Chang'an to Central Asia was long gone.

The Ming Dynasty set up a military fortress at Hami in their early days, but retreated to Jiayuguan at mid Ming, not even controlling today's Dunhuang and Yumen Pass.

Dominance of Maritime Transport

When it came to the 9th century, the Arabs came to Guangzhou, Quanzhou, Ningbo and Yangzhou through the sea. Formerly through the land, only silk was worth long-distance trafficking, but through the sea, although we still use the name Silk Road, silk wasn't the main merchandise. With cheaper ocean shipping, not only silk, but also porcelain as well as other goods, such as rough textiles, handicrafts, ceramics, tea or almost everything, can all be transported. The land Silk Road was in fact of no value in the Tang Dynasty. To the Song Dynasty, there was Western Xia which blocked the route, what was the benefit of going through the land as the maritime transport was sophisticated? Not to mention the Southern Song Dynasty, which only reached Qin Mountains and Huai River with the Jurchen Jin Dynasty in the north. Even in Yuan Dynasty when the route worked, present-day Xinjiang, especially the south part, was under the control of Chagatai Khanate. Admittedly transportation was more developed in the Yuan Dynasty, as both the sea route and land route worked. In addition to Marco Polo, as we all know, the Arabian Ibn Battuta also left records. We can see from their records that they came through different routes, some from the sea, some through the land. This showed that the exchange of people and merchandise was quite developed and diversified, which went beyond the traditional Silk Road. This went backwards in the Ming Dynasty, as it only controlled Jiayuguan in mid-Ming. There were Turpan and other regimes to the west, and ordinary merchants could not even go out of Jiayuguan, limiting the trade mostly within the Western Regions. It was only in Qing Dynasty, not in 1644, but as late as 1759 when Qianlong Emperor crushed the revolt to the South of Tianshan Mountains, that the regime of the Central Plains regained the control of Xinjiang, and finally established a vast unified territory of 13 million square kilometers. But by that time, most trade went through the sea, and the land route was much insignificant.

This is why President Xi called for an understanding of its duration and complexity when talking of Xinjiang issues. In fact, in the 1004 years from An Lushan Rebellion in 755 until 1759, Xinjiang was relatively independent with little control from the central government. We didn't talk much about this period and many people have misunderstandings. It was in this period that Xinjiang converted to Islam. That is why it's so complicated and different. In contrast, although Tibet only came under control as of the middle of 13th century under Yuan Dynasty, it has been kept the same way since the Manchu and Mongolian alliance came into power. Xinjiang is indeed special and complicated in this sense. As we are trying to build the Silk Road economic zone precisely starting from here, these situations should all be considered.

Overall Considerations on Security and National Interests

There was no international order in the ancient times, and equality between nations depended solely on strengths. To make a route work, the dominating country must have proper military and political control over it. It's not that we need to have military control over the route to build the Belt and Road, but the overall security and national interests need to be considered. Stability of the country is still crucial to the building of the Silk Road economic zone.

We must be fully prepared for lurking or existing unfavorable factors and make proper responses. For instance, Central Asia is full of former members of the Soviet Union. Given that they have become independent since the dissolution of the Soviet Union, the influence from Russia is still strong. Although they have their own national languages and are trying to restore their national characteristics, the current official language is still Russian. The upper classes, including intellectuals and politicians, are mostly trained by Russia and the USSR, and many of their leaders have come into power before the collapse of the Soviet Union. Their traditional logistics and personnel relations with Russia still hold. Therefore, when it comes to Central Asia, the influence of Russia cannot be ignored. On the other hand, the United States has always had its global strategic vision. When it took military actions in Afghanistan, it rented a military airport in Kyrgyzstan. NATO's supply in Afghanistan mainly went through Uzbekistan.

From the religious point of view, it's all Islamic from Xinjiang to Turkey; from the language perspective, they all speak Turkic languages. A Uyghur person who goes from Xinjiang to Turkey may understand most of the TV programs, which helps a lot in their communication. Yet we must not forget the two trends since the 20th century: pan-Islamism which considers all Islamic countries as a community of interests, and pan-Turkism that thinks all countries speaking Turkic language should form a community. Some Westerners call this area Turkestan and Xinjiang "Eastern Turkestan". Don't we have the so-called East Turkestan in Xinjiang? In addition to Yaqub Beg's rebellion crushed by Zuo Zongtang, there was the so-called "Three Districts Revolution" in the 1940s instigated by Stalin. The result of such acts was that the three districts, namely Altai at the north of Xinjiang (then called Ashan area), as well as Yili and Tacheng, established an "East Turkistan" Islamic Republic, announced their independence and use Yining as the capital. Later, Chiang Kai-shek tried to make a deal with Stalin and asked him to withdraw the support, and finally the independence movement did not succeed. But even so, in 1962 when China was in dire difficulty, instigated by the Soviet Consulate, tens of thousands of people went across the border to USSR, which made the Yili incident and Tacheng incident. This is the background of some key areas of our economic zone.

In addition, we have conflicts with Central Asia in additional to cooperation. One important contradiction is water allocation. For example, Kazakhstan is short of water, just like Xinjiang. the lower reaches of the Yili River mainly going to Kazakhstan constitute its main source of water, yet Yili River is also an important source for northern Xinjiang. Now China is trying to guide the water to both northern and southern Xinjiang for irrigation. The case is just the opposite for Kyrgyzstan, where the upstream Aksu river passes through. China would also raise objections if Kyrgyzstan wants to build hydropower stations. All these interests need to be coordinated.

In a broader context, we need to consider security for the maritime Silk Road also. Take Malacca Strait, our ideal is to build a route, including railways, highways, oil and gas pipelines, from Pakistan's Guadalan port to Kashgar of Xinjiang, then some of the cargo would no longer need to go by the Malacca Strait. Surely this is good, but if you look at the map, next to this route is Taliban. Pakistan's northern border to Afghanistan is full of refugee camps of Afghans, where Talibanists are impossible to find. Osama bin Laden was hiding in Pakistan. Tribes are strong in Pakistan, where some areas out of the government's control rely on tribal elders. Although Pakistan is China's all-weather friend now, the complex international situation could go out of Pakistan's control. Another big obstacle is the road to southern Xinjiang, which requires constant maintenance. It would be another headache for the future. Or take Myanmar: we had hoped that we could use the harbor of Myanmar and go through oil and gas pipelines and land routes to Yunnan, so that we could open up a new path in the southwest. But we have to pay attention to the political situation in Myanmar. The original investment on hydropower stations has been on hold—can it be restored? Can we continue mining copper there? Can the safety of established oil and gas pipelines be ensured? Can we control it? The former President of Sri Lanka gave President Xi such a warm welcome that China promised to invest USD15 billion to build a port, then the new president came and stopped it. Although he promised that this was not against China, but the variable is always there. To build the Maritime Silk Road, places like the Suez Canal and Panama Canal could all become our soft spots.

One of the "Five Principles of Peaceful Coexistence" passed by the Bandung Conference is called "Mutual non-interference in each other's internal affairs". I think this was very positive when China proposed this, because at that time China's main threat was that others would interfere in China's internal affairs, and China was struggling to eliminate external impacts. China declared that dual nationality would not be recognized in 1950s, mainly to eliminate the concerns of Chinese in Southeast Asian countries, such as Indonesia. There were rumors that all Chinese there were "Fifth Column", so China declared that they had to either naturalize or retain Chinese nationality, and most people chose to naturalize. Today, if the internal changes in other countries are not good for China's development or even damage our interests, what should China do? Shall we leave it completely or actively influence it? This involves our judgment and on the trends of major international issues and control of the situation.

How can China ensure the safety of oversea investments of our enterprises? Now the Asian Infrastructure Investment Bank would be a good move. Some laymen said some nonsense that this was to challenge the World Bank, or to build a monetary fund led by China. We want to promote our high-speed rail, but have we done it? Not yet. There are many reasons. Vietnam was ready, but had domestic oppositions. Thailand is also hesitant, with their realistic consideration. Do they really need so many trains of such high speed, with only a small population? Also, who pays for it? Laos is certainly to build the rails, but it's neither economical nor safe for China to pay for the whole thing. In this case, it would be much safer to have the Asian investment bank invest in ports and railways, funded by a dozen of countries. If we can bring in the US, then it would be even safer with the majority of the World Bank here. This is a practical need. Another example is that, the US engages private security companies, such as Blackwater, to ensure the safety of personnel and properties. On the contrary, it won't work if China wants to send armed police to other countries. Some of these private guards have just retired or even quitted from troops, yet the company was private. They are similar to soldiers, or even more skilled. This might be a solution.

All in all, facing these problems, we must think ahead on how to ensure the security of the Belt and Road. Our foreign policy, especially some strategies needs to farsighted.

We have had many failures in foreign assistance projects which we should learn a lesson from. Take Tanzania-Zambia Railway, our largest project before the reform and opening up. It was Niger that proposed this railway. Surely, we should support them, but the question was why they wanted it. The fundamental reason was that the cooper in Zambia was exported through ports of South Africa, and this was the best and shortest route. But because of the apartheid, there was an economic sanction and South Africa was no longer viable. Consequently, we had to make a detour of more than 1000 km to Tanzania, and go to the sea through Dar es Salaam, so the main purpose was to help Zambia export their copper. If this condition changed, what's the benefit of this railway, no matter how well it was built? Sure enough, South Africa soon ended apartheid, economic sanctions were lifted, and Zambia's copper went once again through South Africa. How could they make use of this 1000-km railway? For tourism, it was through tropical jungle and there weren't many tourists. Also, it didn't make sense for tourists to take a long detour as it was convenient to go along the coast by air or car. I no conditions, there are not so many tourists, and tourists do not need to bypass so far. I was there in 2006 and learned that the railway only run four slow trains every week, which often got cancelled or delayed. We must be more cautious about such inefficient projects in the future.

China Should not Simply Adopt Western Maritime Thought

I want to eliminate a long-held misunderstanding: now we are accustomed to take the Western maritime thought, which is wrong. Western historians, philosophers and geographers all hold such a view—the ocean, representing wealth, hope and communication, is something man cannot live without. This has influenced Western philosophers, and even Karl Marx said that geographical environments build the characters of people: coastal people are open-minded and visionary, while inland people are usually narrow-minded and conservative. Is this right? The ocean in their mind is mainly the Mediterranean, which is totally inapplicable to China. I didn't know this early on and thought they were right. Over the years, with in-depth study, I have come to the conclusion that we cannot blindly apply the experience from the special conditions of the West to China or other countries. Because there is only one Mediterranean, which is very special—it's almost an inland sea with numerous islands and gulfs, which made sailing convenient and safe even in ancient times. More importantly, a galaxy of ancient civilizations embraced the Mediterranean, Mesopotamia, Babylon, Egypt, Rome, Greece, Asia, Assyria, etc. By sailing along the Mediterranean, you can easily access heterogeneous civilizations actively exchanging. There's only one place in China similar to the Mediterranean—Bohai Bay. Unfortunately, despite similar geographical distances, the surrounding Korea and Japan don't have civilizations different from China. In the history, they had been learning from China on a lower level. Japan had copied the whole system of Tang Dynasty. It turned out that China had no incentive to expand maritime traffic. Therefore, the significance of sea for China is completely different from that of Mediterranean for Europe and North Africa.

It would only be natural that the Chinese holds a different view towards the sea. The Chinese view the world as "within the four seas", which means the sea is the border, and there's nothing beyond. When a Chinese came to the sea, he didn't see hope, future or fortune as the Europeans did, but a dead end. It was Chang'an and Luoyang, the politically and economically developed areas, that attracted ancient Chinese. The sea only contributed fish and salt to the ancient Chinese. Sailing was actually quite developed by the end of the Western Han Dynasty, and was attached greater importance, but then it went backwards. Why? It was useless. On the other hand, China hadn't been much threatened by the sea, and it was only necessary to build a navy when there was a threat. The biggest threat was pirates until the Spaniards, the Portuguese and the Dutch came. Hence the Chinese deemed the sea neither as a threat requiring much defense, nor as a treasure generating much benefit.

Ancient Maritime Silk Road Built by the Arabs

Because of this, the ancient Maritime Silk Road was not built by or controlled by China. Who did it? The Arabs. Based on the interests of trade, they found that the land route was blocked and too expensive, but sailing was feasible as they had leading technologies for that. So, they sailed to Guangzhou, Quanzhou, Ningbo, Yangzhou, etc. In the late Tang Dynasty, there were tens of thousands of Arab businessmen and family members in Guangzhou, forming their own communities, known as "Foreign Square" (Fan Fang), which had their own responsible people, known as "Foreign Lead" (Fan Zhang). We can still see the cemetery of Arabs in Quanzhou, with blessings of God written in Arabic, as well as relics of mosques. In the Southern Song Dynasty, PU Shougeng, the executive in charge of Quanzhou's customs and foreign trade, was found to be a descendant of the Arabs. Were there any Chinese people going the other way to trade there? Not that we know of.

The only thing we now know is that in 751 AD, Gao Xianzhi, leading tens of thousands of troops from the Tang Dynasty, attacked Chach (present-day Tashkent in Uzbekistan), when they unexpectedly encountered army from the Abbasid Dynasty. The Tang army was completely defeated and mostly captured. The captives were sent to Baghdad, where the papermaking techniques of China was disseminated by some craftsmen to the Arabs and later to Europe. There was one secretary in the troop name Du Huan, who stayed in Baghdad for 9 years before returning to Tang Dynasty on an Arabian ship. He wrote his experience in about 900 words, from which we learned about the Arabs now. It was the first time that a Chinese had first-hand description on the Arab and Islam world. Recently, an epitaph was found in Xi'an, which said that a eunuch was sent in the late Tang Dynasty to the Arab world, the Tazi. But those were not normal trade. Granted trading was prosperous in the Tang Dynasty—it was all foreigners coming to China. The governments of Tang and Song dynasty were happy just to collect taxes. There are many wrecks in the South China Sea, but as far as I know, most are foreign ships coming to buy stuff, not China's active export. Even in more recent era, the Spaniards, the Portuguese and the Dutch came to trade in the Philippines, Taiwan and Penghu area, but there was no official trading from China. There were some smugglers or entitled businessmen, but totally incomparable to foreign traders. This is something we need to clarify.

Ancient Chinese Governments Never Focused on Economic Benefits in Foreign Affairs

One more thing we want to clarify—what shall we say about the "Treasure voyage" by Zheng He? Is that a paradigm for us to build the Maritime Silk Road today? Objectively speaking, Zheng He's purpose was neither to trade nor to establish a colony, but to promote the emperor's prestige. After the establishment of the Ming Dynasty, the Emperor Yongle felt that it was a shame that less foreigners were coming to China than in the Yuan Dynasty. He wanted to tell them that a strong Ming Dynasty had been established. Another point, more discreetly, was that Emperor Yongle came into power by deposing his nephew with conspiracy, and he would like to attract foreign tribute through such promotion, thereby reinforcing his political legitimacy. He wanted to be the real king, not only supported domestically, but also worshiped by all the countries. That was why he was willing to send Zheng at any expense, not only once, but for seven times. Once arriving at a country or region, the standard formalities of Zheng He were the following: first blast the gun, convene the local monarch or chief with local people, then read the edict announcing the establishment of the Ming Dynasty and the benefit it would give, and encourage local people to come and tribute. If they conceded, he would give out rewards, like gold, silver or silk. Indeed, there were people coming to China, bringing some local products, including spices, rare animals, etc. There was a giraffe from Africa, and reportedly lions and unicorn (could be an elk). From then on, there were elephants in the palace of the Ming or even Qing Dynasty. We didn't see any trade along the route that Zheng He explored. How could such voyages sustain? It was Liang Qichao who first found out his voyages and acclaimed him on the sailing techniques. Apart from some folk stories, the Ming dynasty, notably the ministers, all opposed such voyages. Once Emperor Xuande wanted to see the records of the voyages, and the ministers panicked and claimed them lost, and secretly destroyed them all. Why? They feared that the Emperor might want to do it again, but the treasury had no money for such activities. If we compare him to Columbus, we'll see that although his fleet was no comparison to Zheng He's and he did it later, he was welcomed by the whole city as a hero when he went back to Spain. They all knew that he brought practical benefits, and found new routes where colonies could be established. Zheng He did not do that, and his techniques might as well been exaggerated. According to Prof. Yang You of Shanghai Jiao Tong University, Arabs had reached all the places that Zheng He arrived at. I have noticed the same thing and that was true. Zheng He himself was an descendant of the Arabs with a family name of "Ma". People in his father's generation had been to Mecca for pilgrimage. In the early years of the Ming Dynasty, there were many Arabs remained in China, and records clearly showed that he brought foreign sailors in his voyages. From the records of Zheng He, he used the concept of "needle route" in the maps which came from the Arabs, and used the stars to navigate as invented by the Arabs. Although the ancient Chinese had compass, it was used only locally and totally useless in the sea. If China wants to build the Maritime Silk Road now, there's a lot of innovation and learnings to do.

A miniature of Zheng He's treasure ship

Historically, most activities in China, whether on the sea or inland, focused on political rather than economic benefits, which made them unsustainable. After Zhang Qian's journey to the Western Regions, the Emperor attacked Dayuan for "heavenly horses" and suffered great loss. Hundreds of thousands of people were annihilated. There were also numerous examples of lavish rewards, which raised the expectations of other countries, and drained our own treasury. Emperor Yang of Sui Dynasty built the canal while attaching Korea, led extravagant life, and wanted to flaunt his wealth to foreigners. He treated them in Luoyang and hung silk on trees, claiming China was rich in silk and nobody would need to pay for food. Who would support such "opening up"? Things like this had happened more than once in history. We did not take the initiative to operate foreign trade, leaving the gains to foreign people, while restricting, heavily taxing or even prohibiting private trading. In the Ming Dynasty, there was a shortage of silver, causing the price to soar. Foreigners came with silver from the New World, which should benefit us. Yet the Ming Dynasty prohibited the trade, and local people in Fujian people established armed smuggling groups. When the officials tried to suppress them, they established bases overseas, and even hired Japanese to rebel. We must learn from these incidents and respect economic laws.

Why should we build the Maritime Silk Road? We have so much excess production capacity of cement, steel, coal, iron, etc., and it would be a huge waste if we close them down. On the contrary, if we can invest in infrastructure overseas, building railways, ports, docks, it would be mutually beneficial. The Asia Infrastructure Investment Bank is a very good thing, and nonsense like "challenge the World Bank" or "raise China's voice" can stop. We should note that in Boao Forum, President Xi Jinping was emphasizing bilateral or multilateral development. We should understand that for the Belt and Road, there isn't much precedence that we can copy from. Only innovation can take us through this unprecedented cause, and bring the benefit of the Belt and Road to China and to the world.

Jianxiong Ge Professor of Fudan University, member of the 12th CPPCC Standing Committee, member of History Branch of Social Science Committee, Ministry of Education, Vice President of Shanghai History Association, Counselor of the Shanghai Municipal Government.

The Belt and Road: Revolution of Economic Geography and the Era of Win-Winism

Angang Hu

How do we grasp the essential characteristics of the Belt and Road Initiative? Here, I use two key terms: First of all, economically speaking, it is a grand-scale revolution of economic geography, a revolution that reshapes economic geography of China and countries along the routes as well as the world as a whole. Second of all, in terms of international relations, it gives birth to an era of Win-winism, which is neither similar to colonialism and imperialism a few centuries ago nor the post-WW2 hegemonism. The Belt and Road initiative is the first-ever non-zero-sum game of mutual-benefit and win-win cooperation.

As for the backdrop of the initiative, why was is propose by the leader of China? And why was is accepted by countries along the routes? The most fundamental reason is that China has reoccurred at the center of the stage of world economy. It is the largest trader in the world and one of the largest trade-partners of over 200 countries and regions, which makes China the biggest stakeholder who shares a common destiny with most countries in the world. By taking the lead in promoting Win-winism, China is bound to greatly propel growth in global economy, trade and investment therefore benefit the whole world. From a long-term point of view and in basic trend, it is a revolution of unprecedented scale in economic geography.

From a Strategic Vision to Initiation

The Belt and Road Initiative is the most important "Chines Strategy" that has a global impact since the 18th NCCPC. First of all, it is a Chinese grand strategy. And if "grand strategy is the art of applying the whole of the national power in the most effective way towards attaining the national aim" (British Field Service Regulations, 1935), then China's grand strategy is the art of applying the whole

A. Hu (✉)
Tsinghua University, Beijing, China

W. Liu (ed.), *China's Belt and Road Initiatives*,
https://doi.org/10.1007/978-981-13-0101-8_2

of the national power in the most effective way towards attaining the Two Centenary Goals. It is a general mobilization of 1.37 billion Chinese people, 770 million professionals and 72 million market entities.

Secondly, it is a Chinese global strategy. The west often asks: what kind of global strategies does China have now that you're the second largest economy in the world? What purposes do they serve and how are they different than the US strategies? The American global strategies are established with "consolidating national security of the USA" as its top priority so it can be seen as a strategy that focuses exclusively on fulfilling that purpose. The TPP (Trans-Pacific Partnership) and TTIP (Transatlantic Trade and Investment Partnership), led by the US, aim to reconstruct international trade and investment systems. Their contents and level of trade liberalization are both far beyond those of the WTO' Multilateral Trade Agreements, which are currently under negotiation. It is a set of new standards for trade designed in accordance with developed economies.

Once formed, TPP and TTIP could effectively marginalize current multilateral trade system. Furthermore, many developing countries including China are excluded from the negotiations on TPP and TTIP. The Belt and Road strategy is also different from the US-led European Recovery Program (a.k.a the Marshall Plan), the latter aimed to resuscitate European economy through aid so as to make Europe a major force and instrument against the USSR while allowing the US to control and occupy the European market more conveniently. So, it was a product of Cold-War ideology, whereas the Belt and Road Initiative aims to realize mutual benefit and win-win cooperation as well as fit China's development into the global picture as a driving force. Benefiting both the developing world and the developed, it is a new mode of international development based on Win-winism.

Thirdly, it is an updated version of both domestic and international overall plan by the CPC, ranging from peaceful development to win-win cooperation as well as from within Chinese borders to all over the world. Not only is it an active move towards making Chinese economy globally compatible but also a visionary step forward in promoting China's status as the center and the leader of global economy. It is a first ever since the founding of PRC, especially the vision of the Belt and Road, which is unprecedented for a country of the south.[1]

We see that it only took the Belt and Road Initiative two years to develop from a proposal to a vision then to actions that have become super-sized regional and global cooperative projects (involving over 60 countries and four billion people)

[1]There was the Group of 77 but it lacked substantial vision and action, rather it was more about "statements". The so-called Group of 77 was an international coalition formed and developed in the developing world's struggles against controlling, exploitation and plundering of super powers. In 1963, during a discussion on trade and development at the 18th General Assembly of the UN, 75 developing countries co-proposed a joint declaration. The coalition was then known as the Group of 75. Later at the first UN Conference on Trade and Development, 77 developing countries and regions issued a joint declaration, marking the birth of the Group of 77. By 1979, number of the Group's members surpassed 120, despite its name. As of January 2012, the Group has a total of 132 members.

that attract world-wide attention and are based on mutual benefit and win-win cooperation as well as common development. In a time marked by slow growth in economy and trade as well as high unemployment rate, China stepped up and rose to the occasion. It was a move much expected and widely accepted.

So, where did the vision come from? And how did a Chinese conception become a call answered by countries along the routes and actual cooperation? How did the reshaping of China's economic geography end up reshaping the world? Here, it is necessary that we briefly review the process, results and influence of this decision.

On September 7th 2013, when giving a speech at the Nazarbayev University in Kazakhstan, president Xi Jinping put forth for the first time the following ideas: jointly building an "economic belt along the Silk Road, starting with work in individual areas and link them up over time to cover the whole region.

On October 3rd that year, when addressing the Indonesian Parliament, president Xi Jinping proposed for the first time to build the Maritime Silk Road of the 21st century in a joint effort and the to establish the Asian infrastructure investment bank in order to promote connectivity and economic integration within the region as well as provide financial support for infrastructure construction in developing countries of the region, including ASEAN nations. That was how the initiative to build the Silk Road Economic Belt and the 21st Century Maritime Silk Road (hereafter referred to as the Belt and Road Initiative or One Belt and One Road) came to be. I'd like to like it China's global initiative.

To that end and on the basis of their special research on Xinjiang, the National Research Institute of Tsinghua University furthered their research on the Belt and, on December 1st, submitted the National Situation Report (Special Edition 2013, Issue 22) titled Building a Strategic Artery for China's Westward Opening-up: Strategic Significance and Paths of the Silk Road Economic Belt. We believe that not only is the Belt a combination of politics and economy, domestic and diplomatic affairs as well as space-time cross-over, but also an upgraded version of its historical precursor. In terms of strategic content, it is an integration of westward opening-up and western development; And it was established by generations of leadership through their strategic planning of national security and economy. We also proposed the "five aspects of connectivity": policy, infrastructure, trade, financial (for example, the proposal of Central Asia Development Bank, settlement in local currency and the role of RMB as an international currency in the region) and people-to-people.

In November 2013, the Belt and Road Initiative was included in the Decision issued by the Third Plenary Session of the 18th Central Committee of the CPC thus became a national strategy. In December that years, at Central Economic Work Conference, General Secretary Xi Jinping stressed the importance of working out a plan for the building of the Belt and Road, especially of infrastructure connectivity.

As arranged by the Central Government, relevant departments set up a team for planning and carried out massive preliminary research for Vision and Actions on Jointly Building Silk Road Economic Belt and 21st-Century Maritime Silk Road. They researched and investigated in both domestic and international fronts while listening to suggestions from all sides including countries along the routes. As one of the Chinese experts, I took part in the advice-seeking seminar, organized by the National Development and Innovation Committee, on the Silk Road Economic Belt and submitted a consultant report titled The Silk Road Economic Belt: Strategic Connotation, Positioning and Realization.[2]

In May 2014, when giving the keynote speech at the Conference on Interaction and Confidence-Building Measures in Asia, President Xi Jinping mentioned again that China would accelerate the building of the Belt and Road in a joint effort with countries along the routes. He also stressed that the AIIB would initiate as soon as possible to boost construction of infrastructure in Asia.

In July that year, during the 6th BRICS Summit, President Xi Jinping put forth for the first time that China would contribute more solutions and wisdom as well as public products for the international community, who are earnest to see what China has to offer.

In December 2014, during the 19th collective study of the CPC Central Committee Political Bureau, Xi Jinping definitively pointed out that when it comes to global economic governance, China will not be a bystander and follower but a participator and leader. And that China will have a bigger voice and contribution in international rule-making so as to preserve and expand our developmental interests. Again, he sent a strong signal that China will actively take part in the global economic governance and play a bigger role as a world leader.

What, then, are Chinese solutions, wisdom, voice and contribution? What kind of public products can China provide for the international community? And how will China provide regional and global public products?

In March 2015, after President Xi Jinping attended and spoke at the Boao Forum for Asia, the Chinese government issued the Vision and Actions on Jointly Building Silk Road Economic Belt and 21st-Century Maritime Silk Road (hereafter referred to as Vision and Actions), stating that the Initiative will enable China to further expand and deepen its opening-up, and to strengthen its mutually beneficial cooperation with countries in Asia, Europe and Africa and the rest of the world. China is committed to shouldering more responsibilities and obligations within its capabilities, and making greater contributions to the peace and development of mankind. President Xi made it clear that the Belt and Road should be jointly built through consultation to meet the interests of all. And that, instead of being sealed, it is open and inclusive. It is not China's solo performance but a chorus of all

[2]The report was later published on the Xinjiang Normal University Journal (Philosophy and Social Sciences Edition) 2014, 2nd Issue, co-written by Hu Angang, Ma Wei and Yan Yilong. We mentioned in particular that to build the Silk Road Economic Belt in a joint effort as well as building a strategic land corridor and upgrading China's Go West strategy we need all directional strategic planning. In terms of strategic structure, we need to put a premium on the Shanghai Cooperation Organization along with other mechanisms. And we need to take steady and escalating steps with trade as the main content and advance in multiple dimensions. Through cooperation in safety and stability, economic and trade development, and public diplomacy, we are to further regional cooperation with Central Asia in the five aspects of connectivity.

countries along the routes. The Belt and Road Initiative does not seek to replace current regional cooperative mechanism and initiatives but, based on existing foundation, aims to align the development strategies of the countries along the routes and promote complementary comparative advantages.

Vision and Actions is China's first foreign investment plan and a strategic plan that combines the "Two Big Situations". Not only will it have a positive short-term impact on Asia, Africa, Europe and the world[3] in terms of economy, but also a long-term and profound effect on Chinese and global economic geography.

In less than two years, Chinese leaders including President Xi Jinping and Premier Li Keqiang successively visited over 20 countries and two regional organizations (ASEAN and EU) while holding bilateral and multilateral meetings with other leaders in an effort to elucidate the profundity and positivity of the Belt and Road Initiative. They reached out to learn and develop strategic cooperation and collaboration with other countries while surveying key projects and during which process had expedited a great many of projects and signed numerous agreements such as the China–Pakistan Economic Corridor; the Joint Statement on Cooperation on the Construction of Joint Eurasian Economic Union and the Silk Road Projects; Cooperation with Kazakhstan's Bright Road Economic Strategy; working with Belorussia on landmark projects of the Belt; Cooperation with the Investment Plan for Europe (total amount: 315 billion euros). They also encouraged Chinese companies to participate in infrastructure projects such as Pan-Europe Transport Network, China–Europe Land-sea Express and the New Eurasian Continental Bridge while welcoming European companies to join the building of the Belt and Road. The two sides also hoped to bring the trade volume between them to one trillion USD by 2020. Each agreement represents gigantic amount of investment and enormous value as well as fruitful outcome that are substantial to both sides.

Essentially, China's Belt and Road Initiative is to advocate Win-winism as well as joint construction and connectivity. Firstly, the connectivity of infrastructure in Asia, Europe, Africa and adjacent oceans. Secondly, establish and enhance connectivity and partnership among countries along the routes. Thirdly, build an all-directional, multi-dimensional and complex network of connectivity. One who wins the hearts of all wins the world. Currently, the initiative has received positive responses and support from over 60 countries along the routes and major international organizations therefore kicked off the China-led Belt and Road construction.

[3]Vision and Actions outsets that complex and profound changes are taking place in the world. The underlying impact of the international financial crisis keeps emerging; the world economy is recovering slowly, and global development is uneven; the international trade and investment landscape and rules for multilateral trade and investment are undergoing major adjustments; and countries still face big challenges to their development. The international financial crisis devastated Asia and Africa, leaving the two continents in a slow-growing status. European and Japanese economies came to a halt. China needs to take on the responsibility as the largest engine for growth of world economy, trade and investment.

From All-Round Opening-Up to Advocating Win-Winism

Why did China's Belt and Road Initiative receive such positive responses from countries along the routes? How is it different from other bilateral or multilateral economic and trade relations? What prominent characteristics does it have? The most important one is Win-winism, which possesses the following features:

First of all, the Belt and Road Initiative provides a fully open new type of international cooperation, which means it's open to all countries whether they're along the routes or not. All countries can contribute to their own and regional prosperity through participating the joint construction. This is different from bilateral or multilateral free trade zone, which are open and equal only internally but are exclusive and unequal externally. The Belt and Road are open and equal to all parties, which makes it inclusive, unlike some coalitions or clubs of several countries or regions, non-members can participate and benefit from it.

Second of all, building of the Belt and Road is a mutually-beneficiary win-win cooperation, which means all parties are in a non-zero-sum game where the principles of mutual benefit and win-win cooperation apply to both trade and investment. All countries and regions, regardless of their size, population, wealth or culture, China is willing to have mutually-beneficiary and win-win cooperation with them while seeking common interest. The cooperation will mainly be in three modes: one, the win-win mode, such as bilateral investment treaties; and two, multiple-win mode, such as, joint operation on exploring a third market so the three sides will share the success, as Premier Li Keqiang suggested in Europe, that Chinese companies and countries in Central and Eastern Europe carrying out capacity cooperation in developing countries. China will purchase some equipment from Europe and exerts its advantages in manufacturing and integration. By improving our technologies and capability of energy-saving and environmental protection, China can provide reliable equipment for local infrastructure construction and industrial development; three, the mutually-beneficiary mode, with multiple measures and achievements benefiting multiple parties, such as the China–Europe Railway Freight through the Eurasian Continental Bridge, which is the product of multinational cooperation.[4]

Thirdly, the key of building the Belt and Road is connectivity, especially of infrastructure. As the famous Chinese saying goes: if you want to get rich, build a road first. China is willing to help countries along the routes improve and construct their infrastructure so as to achieve real connectivity among Asia, Europe and Africa and, on top of that, growth in trade, services and direct investment. We also want to establish and enhance connectivity and partnership among countries along the routes and build an all-directional, multi-dimensional and complex network of connectivity.

[4]The Eurasian Continental Bridge, starting from Lianyungang and Rizhao in China, stretches through Gansu–Lianyungang and Lanzhou–Xinjiang railways and leaves China via Alashankou before reaching Europe in three direction from Central Asia.

Fourthly, building of the Belt and Road is the largest cross-regional cooperation in the world so far. It covers the continents of Asia, Europe and Africa, with the Eastern Asia Economic Circle, which is the most active economy in the world, on one end and Europe, the most developed economy in the world on the other. Developing countries and regions in the Middle East, North Africa, Central and South Asia add up to a huge population and potential market. The Belt and Road include at least five cross-regional corridors, among which three are in the Silk Road Economic Belt: from China to Europe (the Baltic Sea) through Central Asia and Russia; from China to the Persian Gulf and the Mediterranean Sea through Central and West Asia; from China to Southeast Asia, South Asia and the Indian Ocean. And the other two are in the 21st Century Maritime Silk Road: from China's coastal ports to the Indian Ocean then to Europe through the South China Sea; From China's coastal ports to South Pacific, again, through the South China Sea. China's role is the promoter of connectivity of infrastructure among the three continents as well as trade and economic integration.

Fifthly, building the Belt and Road is building the largest mutually-beneficiary and win-win community of common destiny in the world so far, a community that includes over 60 countries and more than 4.4 billion people with varied income levels. As an upper-middle-income country, China can, through financial institutions such as Export-import Bank of China,[5] China Development Bank,[6] the Silk Road Fund (40 billion USD), Asian Infrastructure Investment Bank, New Development Bank and SCO Development Bank, help lower-income and lower-middle income countries explore future mega-size markets and cooperate with high-income countries in developing existing mega-size markets. For example, the EU is China's largest trade partner and China the EU's second largest, both sides can jointly develop a third-party market. In the early days of reforms and opening-up, China received massive ODA, including ADD and loans in yen from World Bank. And those low-interest loans and aid had enabled China to reduce poverty and improve investment environment through construction of infrastructure and eventually brought in huge amount of private investment and FDI. Building of the Belt and Road will, step by step, combine official development aid, development financing and market allocation, with connectivity of infrastructure being

[5]Main duties or the Export-import Bank of China are: expanding import/export of Chinese mechanical and electronic products, complete equipment and high-tech products; prompting competent Chinese enterprises to undertake projects and invest overseas; promoting development of diplomatic relations and international economic cooperation; providing financial services. As of 2014, total assets of the bank were worth 2.3776 trillion RMB, with a 1.7417-trillion-RMB loan balance.

[6]China Development Bank serves major medium and long term developmental strategies of national economy by offering financial services such as medium and long-term credit and loan as well as investment. As of 2014, total assets of the bank were worth 1.0317 trillion RMB, with a 794.2-billion-RMB loan balance.

the top priority, provide basic public products for developing countries along the routes and boost growth in commodity/service trade and investment, propelling development of countries along the routes and the world with China's development.

A freighter at sea

Historically, in merely 30 years of reforms and opening-up, China grew from the world's 29th trading entity to the largest, which fully reflected that, from semi opening up to all-round opening up, and from mildly integrated to deeply integrated with world economy, as well as from establishing an open domestic economic system to establishing an open global economic system, China has been a direct beneficiary of opening-up and economic globalization. In the second decade of the 21st century, Win-winism, which China advocates, was fully reflected in the all-round opening up and building of the Belt and Road. Mutual benefit and win-win cooperation, unlike zero-sum games, are repeatable, sustainable and reliable. It is rooted in the same culture and history that gave birth to philosophies like "a just cause gains great support; an unjust one gains little." As the world's largest trading entity and second largest economy, not only does China put forth the vision of the Belt and Road intentionally and purposefully, but it also has the ability and qualification to provide feasible action plans for countries along the routes.

From Reshaping Chinese Economic Geography to Reshaping Global Economic Geography

Essentially, building of the Belt and Road is a grand-scale, profound, interconnected revolution reshaping economic geography. World Bank's World Development Report 2009: Reshaping Economic Geography put forth for the first time an important developmental idea of the 21st century: reshaping global economic geography according to new economic geography theory, new trade growth theory and new economic growth theory. The report established three fundamental facts of density, distance and division. Density refers to economic production per square kilometer; distance refers to the ease or difficulty for goods, services, labor, capital, information and ideas to traverse space; division refers to restrictions to the flow of goods, capital, people, and ideas among countries and regions. It analyzes the reasons for changes in economic geography and market forces, such as agglomeration, migration, specialization and trade.[7] The report draws the conclusion that changes in economic geography are still the fundamental conditions for economic successes in developing countries and regions therefore should be promoted and encourage.

For China, effectively launching and promoting the revolution of economic geography require: free flow and agglomeration of economies of scale and essential factors. For example, urbanization, development zones, etc.; incessantly improving economic density (of population, output, trade, technologies, etc.); enhancing construction of infrastructure, shortening physical and special distances, lowering cost of transportation and logistics, realizing liberalization of trade, investment and facilitation of services, lowering market access threshold, expanding scale of market and entities (private enterprises, individual business owners), enhancing specialization, promoting integration of urban and rural areas, regions (neighboring countries and regions) as well as globalization; reducing population of extreme poverty, increasing income levels, eliminating contrived divisions, realizing equalization of basic public services, improving standard of living (for example, extending average years of schooling, life expectancy), narrowing disparities of income between urban and rural areas as well as development between different regions.

Fundamentally, China's opening-up is not just an economic revolution but an unprecedented revolution of economic geography. I used the word revolution here because it means "a fundamental change resulting in a qualitative transformation." First of all, the revolution of transportation, especially with super highways and high-speed railways, which grew from zero to the longest in the world. By 2014, the total lengths of super highways and high-speed railways had respectively mounted up to be 112,000 km (92,000 km longer than the total length of the US super highways) and 16,000 (10,000 km were under construction); Secondly,

[7]World Bank: (World Development Report 2009: Reshaping Economic Geography) Chinese version, page 49, 74, 97, Beijing: Tsinghua University Press, 2009.

the revolution of energy, especially with the completion of the world's largest national power grid, which covers the largest land area and supplies for the biggest population, with a transmission of extra-high voltage, long distance and large capacity; Thirdly, the revolution of the internet, China has the largest number of internet-users in the world, with 649 million users connected to the internet by the end of 2014, accounting for 21.6% of the total number of the world or 3.2 times as the number of the US. China also has the largest number of mobile-internet users, with 557 million users connected by the end of 2014, accounting for 32.2% of the 2.4 billion users in the world, making China one of the most successful countries in popularizing the internet and narrowing the digital gap; Fourthly, the revolution of urbanization. China has a total of 750 million urban residents, equivalently twice the entire population of the US and more than the whole population (approximately 500 million) of the EU.

All these revolutions took place in the course of the past 30 years or so. There was huge imbalance at first and then it was spread out and applied before the final convergence, with greatly facilitated regional integration in China and, as a result, disparities of development among different regions had kept shrinking since 2004. It means all areas in China are becoming more and more balanced development-wise. Coefficient of variation of per capital GDP among different regions has reduced from 75.1% in 2004 to 43.5% in 2014, down by 31.6%, much lower than the 1995 levels (55.4%); This has promoted urban-rural integration in China, resulting in continuous narrowing of income-disparity between urban and rural populations. The gap went down from 3.33 times in 2009 to 2.92 in 2014.

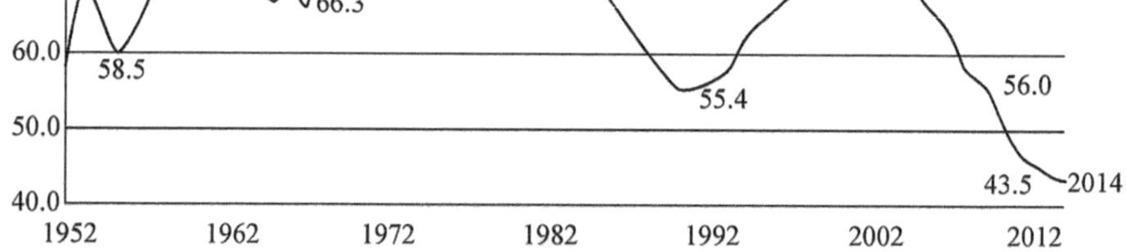

Coefficient of variation of per capital GDP among different regions (1952–2014) (by constant price)

According to China's regional development strategy, coastal areas took the lead and then the development of the western provinces, followed by rejuvenation of industrial bases in the northeast and the rise of central China. The "four plates", so to speak, were reflected in the 11th and the 12th five-year plans as part of the overall planning for regional development and largely reshaped China's economic geography. Since the 18th NCCPC, the central committee of the CPC has successively put forth three development strategies, namely, the Belt and Road Initiative, the coordinated development of Beijing, Tianjin and Hebei Province, and the development of the Yangtze River Economic Belt, thus providing a 2.0 version of overall strategy for regional development, known as the "4 + 3", in the 13th five-year plan as well as highlighting the integration of all areas in China as well as China and the world, so as to reshape China's economic geography and then taking a step forward to reshape economic geography of the world.

Promoting the Belt and Road Initiative, forming a T-shaped geographic pattern consisting of China's domestic pattern (integration of the four plates) and the global pattern, with China's coast-line on the east end and the Eurasian Continental Bridge on the west. Reshaping economic geography of surrounding areas with a dozen of neighboring countries and promoting economic integration within the region have made Chinas positive externality and overflow effect even more prominent. Reshaping economic geography of the world and promoting global economic integration as well as narrowing disparities of development between countries of the south and those of the north. China's Belt and Road Initiative will be a massive and positive driving force.

Connectivity of Infrastructure Is the Top Priority

In the early days of reforms and opening-up, although we didn't know much about the new economic geography theory or the new growth or new trade theories, our leaders and officials at all levels, even average civilians knew that to get rich we must at first build roads, such thinking came to be extensive political and social consensus. It is such simple truth for the modernization of infrastructure is the precondition of modernization of the nation. It is also a truth that's difficult to realize because only an innovative mechanism can accelerate the modernization of infrastructure. And that's why it only took China merely 30 years to become the world's largest market for infrastructure as well as the country that presently has the world's most advanced modern infrastructure, which was the reason behind China's rapid development.

With Vision and Actions, the Chinese government not only proposes a promising vision, but also aims to take solid actions. It's made clear that the connectivity of infrastructure is the top priority in building the Belt and Road. China is willing to help countries along the routes with their planning of infrastructure construction as well as integration of technical standards. We will jointly promote construction of international main corridors so as to gradually form an infrastructure network connecting sub-regions of Asia as well as Asia, Europe and Africa. This is decided by the supply-demand relationship between the two sides and a typical case of Win-winism.

First of all, infrastructure needs upgrading in many developed countries, some of whom along the routes have infrastructure that's so worn out that it's lagging behind the era of high-speed railways. I've visited Australia, the so called "Express Passenger Train" doesn't live up to its name, much slower than the Beijing–Shanghai CRH train that's running at over 300 km per hour. No wonder Financial Reviews (Issue June 30th, 2015) printed an article titled Chinese High Speed Rail should Confine the XPT to History. Currently, the CRH has a total operation mileage of over 160,000 km, with another 10,000 km under construction. Not only does China has the longest and fastest high-speed railway network in the world, but it also has state-of-the-art equipment and engineering. The CRH can run at full speed for eight hours on the Beijing–Guangzhou high-speed rail, which stretches for 2298 km. And it can go for as long as 12 h from Harbin to Shanghai, whereas average foreign high-speed trains can only last for three hours.

Recently, the latest 350-kilometers-per-hour EMU just came off the assembly line and is awaiting the upcoming inspection and optimization.[8] China has put forth the very first edition of National Standards for High Speed Rail Design, which is to be applied to countries who have imported the CRH system, such as Malaysia, Singapore, India, and over 30 other countries that are planning to do so, including Iran, Venezuela and the USA.[9] This has laid a technological foundation for the "Go Global" strategy of Chinese large-equipment manufacturers and provided a set of international standards. During his visit to Europe, Premier Li Keqiang expressed that China has both the capability and the will to play an active role in the investment plan supporting construction of transportation and energy networks in Europe.

Secondly, developing countries have a massive demand of investment in construction of infrastructure. As estimated by Asian Development Bank, Asia's demand for investment in infrastructure is a staggering 700 billion USD annually. In light of this, Mr. Donghyun Park, Chief Economist of the ADB, pointed out that the reason for Asia's faster development comparing to other developing regions is that it managed to secure larger amount of solid investment.

To maintain current economic growth rate, Asia will need more input in the future. This is the case of not only low-income Asian countries, but also medium-income ones including India and Indonesia. Even high-income nations such as South Korea need to increase investment in infrastructure as well. He also commented that so often would foreigners visiting China's for the first time find the country's enormous and high-quality infrastructure impressive. Across China's vast landscape, world-class super highways and high-speed railways stretch in all directions, more are being constructed at record-breaking speed.[10]

---

[8]Economic Daily, July 1st, 2015.

[9]People's Daily Overseas Edition, July 4th, 2015.

[10]Donghyun Park: (Asian Development Bank and Asian Infrastructure Investment Bank can benefit the people of Asia with win-win cooperation), (People's Daily), July 3rd, 2015.

State-Owned Enterprises Strive to Take the Lead and Play the Main Roles

When advancing the Belt and Road projects, China will, as stipulated in Vision and Actions, enhance cooperation with countries along the routes in eight respective fields, among which at least four are closely related to state-owned enterprises, including connectivity of infrastructure; economic cooperation such as in trade and investment; industrial investment; cooperation in energy and resources. To that end, Chinese state-owned enterprises, especially central SOEs, will play a major role.

With rapid strengthening of state-owned enterprises, especially central SOEs, total number of Chinese companies (altogether 100 of them, 92 are mainland companies, 83 are state-owned, 59 are central) in the top 500 has surpassed that of the Japanese ones. As showed by data provide by the SASAC, by the end of 2014, a total of 107 central SOEs had established 8515 overseas branches in over 150 countries and regions. Since the 12th five-year plan, central SOEs' total overseas assets value has grown from 2.7 trillion yuan to 4.9 (equivalently 800 billion USD, more than twice the GDP of Thailand), with an annual growth of 16.4%; And their operating revenue has grown from 2.9 trillion yuan to 4.6 (equivalently 750 billion USD), with an annual growth of 12.2%.

Overseas investment of Central SOEs accounts for approximately 70% of China's entire OFDI in non-financial sectors. Also, they account approximately 60% of the total operating revenue of all foreign projects contracted to China. Central SOEs have achieved prominent successes in promoting international endeavors of Chinese industries such as electric power, rail transportation and construction materials.

China has become the world's second largest contractor of foreign projects, with the total contract value growing by 42.7%, from 134.4 billion USD in 2010 to 191.8 billion USD in 2014. Value of completed projects grew by 54.4%, from 92.2 billion USD in 2010 to 142.4 billion USD in 2014. During January to May 2015, new foreign projects contracted to China were worth 67.58 billion USD totally (equivalent to 414.27 billion RMB), a 26.6% year-on-year growth; Accomplished turnover was 51.63 billion USD (equivalent to 316.49 billion RMB), a 10.1% year-on-year growth. Foreign project contracting operations have propelled internationalization of Chinese enterprises and accelerated the "Go Global" process of China's equipment manufacturing industry. In 2014, a total of 62 Chinese contractors were listed in the world top 250, with a total turnover of international projects (79.135 billion USD) slightly lower than the combined turnover (79.436 billion USD) of 13 Spanish contractors on the list and higher than that (70.956 billion USD)of the 31 American contractors. There is one Chinese contractor in the top ten and five in the top 30 (see Table 1).

Take China Communications Construction Corp for an example, in 2014 the company's turnover of international projects reached 13.2 billion USD, putting the company at 9th place in the world and 11th among the 150 largest design firms. The company was contracted to build five of the ten largest container ports in the

Table 1 Ranking of Chinese companies in the top 250 international contractors and top 500 companies (2014)

Company	Place among top 250 international contractors	Operating revenue from foreign projects (hundred million USD)	Percentage in the world total (%)	Place among top 500 companies 2014	Operating revenue (hundred million USD)
China Communications Construction Corp.	9	131.62	24.3	187	546.1
China State Construction Engineering Corp.	20	57.43	5.9	52	1108.12
Power Construction Corporation of China	14	53.14	25.7	313	431.60
Sino Mach	25	52.89		278	394.18
China Railway Group Limited	28	47.67	5.4	86	911.53
China Railway Construction Corporation Ltd.	39	34.86	3.6	80	957.47

ENR: The top 250 international contractors

world, five of the ten largest cable-stayed bridges in the world, four of the ten largest suspension bridges in the world and five of the ten largest sea-crossing bridges in the world, all of which are of the highest engineering level. Among them the Hong Kong–Zhuhai–Macau bridge is a gigantic bridge connecting the three cities. The first phase of construction is to be completed in 2017. Total cost of the bridge is over 100 billion yuan and the construction will take eight years to finish. It is a super project that receives world-wide attention and a super tough one that is designed to last for 100 years. The company was also contracted to build the Gwadar Port, a key project of the China–Pakistan Economic Corridor. Another engineering achievement of the company is the Colombo Port, which has earned the name "the Pearl of South Asia on the 21st Century Maritime Silk Road" and cost 1.5 billion USD in the first phase of construction. Furthermore, the company was contracted to build the second sea-crossing bridge of Penang, Malaysia, which will be the longest of its kind in Southeast Asia.

In China Railway Group's case, the company is now one of the largest contractors of construction projects in the world, ranked 85th among the top 500 Companies in 2014 and 28th among the top 250 contractors. Almost all highway-railway bridges and large railway brides on the Yangtze River were built by the company, who currently is building another 12 bridges on the River.

They have also built the toughest and the most magnificent Qinghai–Tibet Railway and bridges. Besides domestic projects, the company also builds brides in Southeast Asia, South Asia and Africa. For example, the company designed and built Padma Bridge, the largest one in Bangladesh, along with other projects in Myanmar and Bangladesh. In 2014 alone, the company signed a total of 43.2 billion yuan (7 billion USD) worth of contracts and accomplished a turnover of 30.3 billion yuan (4.9 billion USD).[11] Judging by number and quality of bridges, China Railway Group is the best among all Chinese companies.

Another example is China Railway Construction Corporation, who had an operation turnover of 3.486 billion USD in 2014, ranked 39th in the world. Recently, the company has just signed some new contracts of foreign projects worth 127.8027 billion yuan (equivalent to 20.8 billion USD), among which, the coastal railway project with Nigeria along worth 11.97 billion USD. Not only is it the highest-valued foreign project contract China has signed so far, but the entire system is going to adopt China's homegrown technical standards for railway as well.

As for State Grid Corporation of China, they set up a team, as early as November 2014, in charge of connectivity with neighboring countries to accelerate preliminary procedures. The company has built various kinds of power facilities based on coal, wind and solar energy while exporting electricity to foreign countries. Currently, while making breakthroughs in new markets such as countries along the Belt and Road, the company is actively carrying out cooperation and negotiation on many projects so as to improve their international reputation as well as gaining more international market share and helping Chinese power equipment "go global".

These SOEs have demonstrated in global market the collective image of Chinese contractors, Chinese engineering, Chinese manufacturing and Chinese brands.

So, in the future, what can SOES, especially central SOEs, do better when "going global" and how will they play the leading role as the main force?

Firstly, they will implement all requirements as stipulated in Vision and Actions, which are to serve as the guideline and foundation of the Belt and Road strategic planning.

Secondly, they will, through due diligence and elaborate planning, determine strategic goals and directions so that they can choose key fields, countries and regions among others as well as select important projects and raise funds.

Thirdly, they will innovate implementation of "going global" and improve level of international cooperation through flexible practices of trade, investment, construction of industrial parks, technical cooperation and talent training, etc.

Fourthly, they will innovate modes of overseas operations, such as joint ventures and combined actions. They will also carry out a variety of flexible domestic cooperation such as that between upstream and downstream as well as of international capacity, equipment manufacturing, international project contracting and service trade.

---

[11]Economic Daily, July 6th, 2015.

Fifthly, they will enhance risk prevention and control to ensure stability of operation and safety of assets.

Sixthly, they will accelerate training and recruiting of talents with international experiences. The biggest barrier preventing Chinese companies from becoming international competitors and global giants is not capital but talents, especially experts and managers with global perspective and experiences. To that end, they need to send Chinese talents to study and be trained overseas so they can be internationalized. Also, they need to attract international talents with recruitment plans. Those are the two engines that will drive Chinese companies, along with their R&D levels, towards internationalization.

Finally, they must balance domestic and international development as they explore both markets while utilizing resources and innovate technologies. As showed by data provide by the SASAC, central SOEs invested 318.8 billion yuan (equivalent to 51 billion USD) in research and development in 2013, not only did the amount account for over one third of national total, with a 20% annual growth, but it's also higher than that of an average developed country (for example, the UK with 46.3 billion USD and Italy with 45.6 billion USD). Together, central SOEs own a total of 610 national R&D facilities, about one fourth of national total, and one fourth of China Patent Gold Award winners. A great many of major innovative achievements such as manned space exploration, the lunar exploration project, high-speed railway, 4G mobile communication, the Beidou Navigation Satellite System, extra-high-voltage and smart power grid, oriented silicon steel, million-ton level coal liquefaction and third-generation nuclear power plants are turning Chinese technologies into world technologies, Chinese standards global standards, and made in China made with the world.

To summarize, when building the Belt and Road, state-owned enterprises, especially central SOEs, should become China's most powerful weapon and instrument as they collaborate with flexible competitive civilian companies and other market entities to form a collective Chinese legion in joint operations and building of Chinese brands.

Staring a New Era of Win-Winism

Historically, economic globalization of the world went through the following stages: colonialism, occurring some 500 years ago; imperialism, 200 years ago, including Fascism, which was the source of two world wars; post-war Hegemonism over the last 70 years, marked by the rivalry between the US and the Soviet Union, followed by the US supremacy, which was the very source of the wars in Iraq, Afghanistan, Libya and Syria.

President Xi Jinping is an advocator of Win-winsim: "Peace instead of war, cooperation instead confrontation, win-winism instead of zero-sum game, are the eternal themes of peace, progress and development of human society."[12] Led by China, the world will embrace the era of Win-winism in the 21st century.

Compared to the previous times, the most distinctive features of the era of Win-winism are as follows: instead of being unjust, this era is just; instead of being discriminative, this era is equal, instead of being exclusive, this era is inclusive; instead of being confrontational, this era is non-confrontational, instead of being disruptive, this era is harmonious; instead of being unsustainable and temporary, this era is sustainable and lasting.

China's Belt and Road Initiative symbolizes the beginning of the era of Win-winism and will enable China to play a bigger role as well as make more contribution in terms of peaceful and win-win development of the world.

Today, unlike any other time in history, China is right at the center stage of world economy and has become the largest stakeholder to all countries of the world. Never before has China played a bigger role in world economy by being the most powerful engine of growth in economy, trade and investment. Nor has China had a bigger influence on reshaping economic geography or been a stronger force in promoting connectivity of infrastructure globally.

More and more China needs the world, and vice versa. All countries, big or small, wealthy or poor, in Asia, Europe, the Middle East or African, are paying close attention on China's future and listening to China's voice.

The most creditable parts of the Belt and Road Initiative is its overflowing and positive externality, which greatly mobilized the collective rise of countries of the south and fundamentally changed the economic trend between the south and the north in the past two centuries. Countries of the south have become the "main force"[13] in the 21st century with the south and the north in common prosperity. China has both the will and the ability to help countries of the south rise collectively while kicking off an era of Win-winism as well as building a harmonious, commonly prosperous and green world.

---

[12] Xi Jinping: Remember the History, Explore the Future, Rossiyskaya Gazeta, May 7th, 2015.

[13] As of 2014, China's population accounted for 22.6% of that of countries of the south, and its GDP(PPP), 28.7%, goods and service export, 27.8%.

Angang Hu Dean, the National Research Institute of Tsinghua University Professor, Doctoral Supervisor School of Public Policy & Management, Tsinghua University; Member, Committee of Experts for the 11th, 12th and 13th Five Year Plan; Member, Chinese Economists 50 Forum; Mr. Hu has been working in national research for over 30 years and is one of the pioneers in the field.

The Belt and Road and a New Model of International Development Cooperation

Yifu Lin

Introduced by President Xi Jinping in 2013, the Belt and Road is China's new initiative on international business and foreign affairs, in order to promote mutual political trust, economic integration and cultural inclusion among countries along the Belt and Road by enhancing connectivity, infrastructure and trade. For many of those Asian and European countries, the low economic development levels and backwards infrastructures are impeding their economic growth. This new type of international development corporation, in the short-term, can create jobs and stimulate growth for them, and improve their economic growth potential in the long run. This changes the development assistance by "blood transfusion" from developed countries to the new development cooperation mode of "blood generation". The infrastructure construction in developing countries also contributes to global economic recovery, which benefits both the developing and developed world, making it a new win-win mode of international development cooperation.

In February 2009, as Chief Economist and Senior Vice President of the World Bank, I faced the challenges of the international financial crisis. In order to propel global economic recovery, I advocated the "New Marshall Plan" with infrastructure development as the pillar. Although the starting point of the Belt and Road initiative may be different, we both come to infrastructure as the starting point. I had done a detailed analysis on the benefits of infrastructure construction for both developing and developed countries in the "New Marshall Plan".

In September 2008, the collapse of Lehman Brothers collapse triggered the worst global financial crisis since the Great Depression of the 1930s. Developed countries, the fuse of this crisis, need structural reforms to fully recover, but this would mean reducing wages and benefits, deleveraging financial institutions and reducing fiscal deficit, which would suppress consumption and expand unemployment in the short term. This would be politically difficult at this point of post-crisis high unemployment. Past crises broke out mostly in developing countries or regions,

Y. Lin (✉)
Peking University, Beijing, China

W. Liu (ed.), *China's Belt and Road Initiatives*,
https://doi.org/10.1007/978-981-13-0101-8_3

where IMF would advise expanding exports by devaluation to counter the contraction of structural reform and make room for the reform. On the contrary, this crisis have tended simultaneously in many developed countries, which all have the need of structural reforms. If one country wants to increase exports by devaluation, a devaluation race from other developed countries would ensue. No one can increase exports in the end and the structural reform would be tough. Foreseeing the failure of traditional IMF actions, I proposed a "New Marshall Plan" of global recovery in early 2009, mainly advocating intensive investment on infrastructure across the globe, particularly in developing countries. These infrastructure investments not only will eliminate bottlenecks and drive economic growth of the receiving country, but also drives the exports of other countries, including developed countries, creating room for structural reforms just as devaluation would. It is a win-win strategy benefiting both developed and developing countries, for now and for the future.

The reason that I proposed this plan, is that not only the infrastructure in developing countries other than China are very poor, but also the developed countries can take this chance to improve. For developed countries, rather than doling out unemployment benefits at the expense of budget deficit, investment in infrastructure development in themselves and developing countries would create demand and jobs in the short term, and improve growth potential and increase tax revenue to make up the deficit in the long run. In addition, the yields of sovereign funds, pension funds, China and the massive amount of government bonds bought by China and some oil exporting countries are so low that infrastructure investment would make more sense.

Since proposed, this initiative has gained increasing popularity. The 2010 G20 Summit in Seoul listed infrastructure as the No. 1 key pillar for developing countries in the development consensus, while the 2013 summit in St. Petersburg and 2014 summit in Brisbane reemphasized the importance of infrastructure. The 2014 "World Economic Outlook" by International Monetary Fund also advocated infrastructure as a measure of economic recovery, but until now no country has taken action to set up global recovery fund for infrastructure. Now China has put forward the Belt and Road initiative and established the Asian Infrastructure Investment Bank (AIIB) and the Silk Road Fund, setting up the example by leading the infrastructure development of Asian and European countries. This is a win-win initiatives benefiting both developed and developing countries, and we welcome all countries, including Europe, US and Japan to join us.

The Belt and Road initiative propels the economic recovery and development globally, particularly in the surrounding developing countries in Asia and Europe. During the development, it stimulates local demand, creates jobs and boosts economic growth, and after the completion, it eliminates the bottlenecks of growth and "generates blood" for both the constructing country and those connected, enhancing the potential for subsequent development. This initiative is also beneficial to China—it creates more use of foreign exchange reserves in addition to building a friendly external environment. Growth in countries along the Belt and Road would give more potential to China's export growth. The initiative may also help China get the resources it needs from resource-rich countries.

China has another trump card to drive this new cooperation model of "blood generation"—its gigantic labor-intensive industries. As the wages increase, China's comparative advantage is disappearing in this respect, and a big part of such industries will be gradually shifted elsewhere. Such shifts have been witnessed repeatedly since the industrial revolution, creating opportunities for the receiving countries to industrialize and modernize. Countries seizing the opportunity would have rapid development of 20–30 years, freeing themselves from poverty and becoming new industrialized economies.

A new feature of this shift of labor-intensive processing industry is a massive scale. According to the third industrial census in 2014, China's employment in manufacturing is around 124 million people, 12 times of that of Japan when it did the shift in the 1960s, and 22 times of the "Four Asian Tigers" in the 1980s. The only place to undertake labor-intensive industries of such scale is Africa, with a billion people, large number of surplus labor force in rural areas, high proportion of young people, and wages only one-tenth to one-fifth of workers in China. China has its success stories to do such shift—In 2012, Dongguan Huajian Footwear set up factories in Ethiopia with both the supply chain and the target market there, creating 2000 jobs and making profit in the same year, doubling Ethiopia's footwear exports, and becoming the largest exporting enterprise in Ethiopia.

By shifting labor-intensive processing industries from China to Africa and other developing countries, not only would it drive the GDP and GNP up, but also helps with the export of high-added value intermediate products and equipment, expanding overseas markets for China's industrial upgrade. Therefore, this would be another win-win model of development cooperation.

The primary purpose of the Belt and Road initiative is to enhance the infrastructure and improve connectivity among Asian and European countries, supplemented by economic and trade cooperation. As for the shift of labor-intensive processing industries, the strategic goal of "Community of Shared Future between China and Africa" proposed by President Xi Jinping during his first visit to Africa after he became president in 2013 would be our goal—focusing on the shift of labor-intensive industries, supplemented by infrastructure development. Both strategies would "generate blood" for developing countries cooperating with China.

This new development cooperation model focusing on infrastructure development and industrial shifts is of great significance to international development. Since World War II, developed countries have setup up numerous multilateral or bilateral development agencies, such as the UN Development Agency, the World Bank, USAID and DFID, yet not have not been able to truly help the developing countries, of which the vast majority are still struggling in the low-income and middle-income trap. Since World War II till now, among close to 200 developing countries and territories, South Korea and Taiwan are the only two rising from low-income to high income, and only 13 economies went from middle-income to high-income, of which 8 are close to Western Europe, and the other 5 Japan and the "Four Asian Tigers". Why the assistance from developed countries has not been up to our expectations? The developed countries don't have comparative advantages in labor-intensive industries and infrastructure construction, as they are already over those stages. So they can only do "blood transfusion" to developing countries, with many additional conditions not directly related to the development. With the Belt and Road and the "Community of Shared Future between China and Africa", China can help eliminate the bottlenecks to growth in developing countries, provide opportunities and introduce "blood generation". All developing countries can enjoy real growth, realizing the vision of "flourishing spring garden" depicted by President Xi Jinping.

Granted, China faces many challenges in implementing the new cooperation model of foreign development. On one hand, Chinese enterprises are not really prepared to "go out". This strategy of "going out" was only promulgated in 2001, when the political, financial, and legal environment were less favorable, both domestically and internationally. Many enterprises went out but ultimately failed. This requires learning on the enterprises' part, but also help from the government help. On the other hand, China's foreign cooperation policies come from various authorities—the Ministry of Commerce, Ministry of Foreign affairs, Ministry of Finance and Ministry of Agriculture, among others, caused some confusion. China also set up numerous funds for foreign aid, such as the Export-Import Bank of China, the Development Bank, the Sino-African Fund, Silk Road Fund, dispersing much of the strength. It would be necessary to learn from the experiences of developed countries, establish a centralized international cooperation ministry to coordinate international relations and synthesize resources, as to bring China's international development cooperation to a new level.

A caravan of camels in the desert

Yifu Lin graduated from University of Chicago with Ph.D. in Economics, and did post-doctoral research at Yale University. He has been members of the 7th, 8th and 9th National Committee of CPPCC. He is the deputy director of the economic committee of CPPCC, and vice-chairman of All-China Federation of Industry and Commerce. He was formerly Chief Economist and Senior Vice President of the World Bank, and is now the director emeritus of the National School of Development at Peking University, and an advisor of the State Council.

Grand Strategic Thoughts with a Global Vision

Liang Qiao

As a descendant of war veterans and a contemporary military personnel of China, I'm very honored to visit Zhijiang on the 70th anniversary of the victory of the War of Resistance Against Japanese Aggression. I would like to pay my tribute to the honorable city and the people of Zhijiang.

70 years ago, at this place, the Chinese army successfully resisted the attach of 100,000 Japanese soldiers. Zhijiang battle was the last battle in the War of Resistance, and it was a complete victory. Zhijiang became the first city to accept the Japanese surrender. This was a moment of pride for the Chinese people.

70 years later, China has undergone tremendous changes. All these developments are closely related to the victory 70 years ago. If we could not overcome the invaders who invaded our country twice, we couldn't have come to this today. Today we commemorate the martyrs, and look into the future with full confidence in China.

Who Is Driving the Decline of the Financial Empire?

Today, whether we talk about the international situation, or talk about China's situation, two most important factors are unavoidable: the rise of China, and the decline of the United States. There are a lot of people who do not agree with the decline of the United States, saying I'm too pessimistic about the US, and I should go to the United States to see the "recovering economy" without a sign of decline. But what I meant by the decline is not for the US we are seeing today. There is no doubt that the United States is still the strongest country in the world, the No. 1 in nuclear or conventional military power, the No. 1 in economic aggregate, the No. 1 in technological innovation, all these titles they still hold. Yet undeniably our

L. Qiao (✉)
Peking University, Beijing, China

W. Liu (ed.), *China's Belt and Road Initiatives*,
https://doi.org/10.1007/978-981-13-0101-8_4

August 21st, 1945, Zhijiang Airport, General Takeo Imai, who surrendered to China on behalf of Japan, rode with his retinue in a jeep with white flag as it went around the field for demonstration. Courtesy of xinhua.net

times has undergone profound changes, which started from the US. The most important one was that the Americans invented the Internet and brought it to the world. The popularization of the Internet has fundamentally changed the world, including the US. I have discussed with some American scholars, and I said, if you thought that China was the most powerful challenger to the United States, you were wrong. China is taking its own way, rather than trying to challenge the United States. What really challenges the future of the United States, especially its leadership in the world, is the US itself.

It is the American "innovation" that brought itself on to the path of decline. Americans have always proudly declared: "The US is the No. 1 innovator." Indeed, the scientific and technological innovation of the US is leading the world, but it is exactly that that put the US on the track of inevitable decline. Of all the innovations that the U.S. takes pride in, the most important one is the Internet. Since its birth, the Internet has been a multiplier to industrial production, the economy and military innovation. However, when Internet reaches all over the global, some irreversible changes would happen, which is its most essential features: decentralization.

Why would "decentralization" lead to the decline of the United States? Because it will deconstruct power. When the Internet is extremely popular, the trend of "decentralization" and "multi-center" would show despite of anybody's will. This trend itself will inevitably lead to the multi-polarization pursued by all the countries, and thereby deconstruct the super-hegemony of the US. Obviously, the US is not yet fully aware of this, but it already has a fear of their own unescapable decline. Under the influence of this fear, the United States did not cope with the trend with ration and make effective adjustment, but repeated the history of all empires in decline: they thought the hegemony can be retained as long as the challengers are suppressed, and make the 21st century another "American century". This grand strategic misjudgment leads the United States to select, or locked in China. For China, this is certainly not a good thing, but not entirely bad either. A Chinese saying goes "a person without pressure is frivolous". When the suppression from US becomes a pressure, China might have a stronger come-back. Newton's law has told us that, for every action there is an equal and opposite re-action.

The Fundamental Competition Is for Capital

In a period of economic transition, the economy of China will inevitably go down, and this trend has become the new norm of China. In the meantime, the economic recovery of the US is also weak. Predictions of a strong recovery have been around since the beginning of 2014, and especially in the fourth quarter of 2014, the US economic grew by 4%, taking the whole world by surprise. But just a few months later, statistics of the first quarter of 2015 showed the US economy has slowed down to 2%. What's the reason behind this roller-coaster ride? Probably only the American would know the secret.

Economists like to say that the modern economy is the economy of confidence. When a country's economic data is not good, investors around the world will lower or lose confidence in its outlook, driving the economy further down the spiral. The Americans understand this better than anybody else, as its economy is relying on debt. It needs good economic data badly to attract international capital back to the United States. This means that the US needs to work hard on the data, and every data that shows a "strong recovery" of its economy might mean more than it appears. Nowadays, the fight for international capital more intensive than ever. Between the United States and Europe, between the United States and China or other regions, the data war has become an important approach.

The Ukrainian incident is a typical battle for capital between the US and Europe. Around China, the 2012 dispute over Diaoyu Islands, the Ocean Oil "981" platform and Occupy Central incident in Hong Kong ... all have their political and military factors. But what is really behind those political and military factors? National interests, which ultimately can be demonstrated by capital. So, the competition for capital is the most fundamental one.

Why China suddenly faces so many disputes? Because China has come to the point of becoming the second largest GDP. In 2014, the world's total GDP reached 77 trillion USD, where 17 trillion USD from the US, 10 trillion USD from China, and more than 4 trillion from Japan. China surpassed Japan four years ago, and took a big lead just in a few years, becoming the undisputed second in the world.

In this situation, some collisions between China and the United States seem inevitable. The United States is prepared: it proposed some new strategies, such as "Asia-Pacific rebalance" in 2010 targeting China. It shifted its strategic focus to the Western Pacific, moving 60% of the Navy to the Pacific. This showed that the United States was taking China as the main opponent. In addition to political and military pressures on China, the United States tried to impose economic pressure through the "TPP" (Trans-Pacific Partnership Agreement). As an economic strategy to alienate China, TPP has three goals. The first one is for the economic security of the United States. They believe that in the past 20 years, the US economy is too dependent on China, and something else is needed to get rid of this dependence. The second one is to set up new rules for the Internet economy. The rules of trade of the World Trade Organization (WTO) were established under the dominance of the US 17 years ago, and many of them are outdated in today's global economy. At the same time, the emergence of the Internet economy calls for new rules, while the US does not want China to set the rules. It tries to exclude China once again, and only accepts China when the rules are fixed, and force China to make more concessions like it did when joining WTO. The third one is to block China, Japan and Korea from setting up a free trade zone in Northeast Asia. Why does the US hate it so much? It learned the lesson from the European Union and Euro. The European Community has become the EU, and successfully launched Euro, making EU the world's largest economy overnight, only then the US found that the hegemony of US dollar was challenged. No more new challengers this time.

Northeast Asian Free Trade Area, once established, would become the world's third largest economy. And it won't stop: it will surely go to Southeast Asia and become East Asia Free Trade Area, then to the west to cover South Asian sub-continent: India, Bangladesh, Sri Lanka, followed by the integration of Central Asia and West Asia, forming an economy of 50 trillion USD, even more than EU and North America added up. Will such an economy be willing to use Dollar or Euro to settle the trade? Certainly not. It will launch its own currency, and there will be three major currencies in the world: the dollar, the euro, the yuan. The US can only have one of it. Is one-third of a hegemony still a hegemony? Having learned the lesson of the EU and Euro, the US is bound to take preventive measures. As a result, through the Sino-Japanese Diaoyu Islands dispute, the US successfully blocked the negotiation process of the Northeast Asian Free Trade Area, and nipped the buds of the Asian economic community or the potential regional currency. This is how we should understand the situation from a global perspective.

Neither Arms Race, Nor Self-weakening

In order to facilitate the eastward shift of strategic focus, the US surely would take some military moves. That was the military strategy launched by the Pentagon in 2010—AirSea Battle, targeting China and Iran. In fact, a medium-sized country like Iran would not amount for a strategic opponent, so the main target would surely be China. But even for such a concept, the US is aware of its own weakness. In 1996, China trialed missiles in the Taiwan Strait to warn the separatist forces in Taiwan. In 1999, the Chinese Embassy in Yugoslavia was bombed. These two incidents made the Chinese government aware of the importance of military force while developing the economy. Since then, the investment on military forces have been expanded as allowed by the budget, yielding great progress. This in turn has become a concern of the Pentagon.

In this Air-Sea Battle strategy, the U.S. believe that that there won't be a war between China and the U.S. within 10 years. It is not that they really cherish peace, but after studying China's military power, the U.S. realize that the current capacity of the U.S. army would not be enough to offset some of the advantages that China has established, like the capacity to attack aircraft carriers and to destroy space systems. Therefore, the United States must take ten years to develop a more advanced combat system to offset some of China's key advantages. This means that the possible war schedule has been pushed back for 10 years. Although the war may still not happen even after 10 years, we still have to be on alert all the time. If China wants to make sure that the war will not happen 10 years later, we would need arm ourselves better, and be prepared for the war.

One of the important reasons of the collapse of the Soviet Union is military. The US dragged the USSR into the mire of "Star Wars", draining its limited foreign-exchange reserves and causing great difficulties to his economic development. Clearly the U.S. is hoping China to follow the same path through AirSea Battle, yet we won't. China should not be in an arms race with the US, yet cannot give up its own power. The only choice is to find a way of military development that suits China. This is also part of the context for China's strategic choices.

Significance of the Belt and Road

Historically, in the course of the rise of all the great powers, there has been globalization around them. Globalization is not a consistent process from the ancient time till today, but a piecewise process for each era. Whether it was the ancient Rome or the Qin Empire, the "globalization", in today's view, would only be regional imperial expansions. The real globalization of the modern history started with the British Empire, which was the globalization of trade. Inheriting the British, the US continued the globalization on trade, but the real one of its own is the globalization of US Dollar, which is the one we are experience now. Indeed, after World War II, the US did not occupy any country's territory, yet brought everybody into its global financial system with the US dollar. Americans may not be on your soil, but the US dollar is in your market and drives your economy. All the countries are producing for dollars, and all products and all assets are valued in dollars. This is a financial colonization, a smarter colonialism than that of any colonial empire in history. In this sense, I do not agree that the Belt and Road is in line with "global economic integration", which is tantamount to saying that the Belt and Road is going with the globalization of dollar. At this point when the US accused China of "free ride" and tried to exclude China in TPP, this kind of understanding is clearly off-track.

The strategic vision of the Belt and Road is a choice of path that China made, trying to find new directions of economic development, but also a countermeasure to the pressure from the US. It can be seen as a counterbalance to the shift of strategic focus and Asia-Pacific Re-balance. Some people might say that, a countermeasure should go head-in-head. How can you counterbalance by turning around? That's very true, and that's exactly the essence of the strategy. By avoiding head-to-head conflict and extending towards the west, it can alleviate the pressure from the US. Therefore, we should have a more in-depth and a comprehensive assessment on the strategy of the Belt and Road. Different from the empires choosing zero-sum games or winner-take-all, China chose a path of cooperative games, solving not only the problems of China, but also of other countries through cooperation and mutually beneficial actions. This would minimize the friction with the US in a larger geopolitical area.

How to Move the Belt and Road Forward?

The "dream for blue" of the Chinese soared since a popular feature film was on air. Since then, there has been a lingering illusion, that a country would not be a power if it is not a maritime power. In fact, many powerful countries in history are not maritime powers. Today, China will become a maritime power sooner or later. In the future, 70% of their resources and energy would come from the sea. Countries without the power to get these resources would be struggling to survive. Therefore, it's not an option for China not becoming one. But what we need most now is the opportunity for the next decade. "Maritime Power" will not solve the imminent problems. A navy cannot be built over 10 years, but China cannot wait for another 10 years.

That is why China chose the Belt and Road as a new national strategy. Though seems to bifurcate on both the land route and the sea route, the Belt and Road, in my view, is more about the land than the sea. At first glance, the sea route is easier and more attractive, as long as you are not in a war and the routes unblocked. Goods coming out from ports will arrive in around 10 days, and it's much cheaper than land transport. In peacetime, naval escort would suffice to cope with pirates and ensure the safety of maritime transport.

However, when international relationships, especially those between powerful countries, deteriorates or even endangers the survival of a country, maritime transport would be hazardous. For a big country like China, there are only a few opponents, or maybe only the US that can endanger it on the sea. With its power, the US can cut off the maritime routes if things go bad. With such a possibility, the importance of the "belt" (Silk Road economic belt) becomes more pronounced. The "belt" break through the places where the US hasn't influenced much. If the US wants to block China on the "belt", it would need to negotiate with each country. Currently, the US only reached an agreement of using the military airports with one country, without anything else with other Central Asian countries, who are vigilant about the intervention of the US. It would be extremely costly for the United States to instigate each one of them against China. Buried in debt itself, how could the US fool others? Most countries are profit-chasing—why would they help you block China if there's no benefit? The Chinese comes to bring them benefits, while the U. S. cannot. Without actual benefits, the U.S. cannot expect other countries to work with it to suppress China, just because it doesn't like China, or the ideology or political system of China, or sees China as a challenger. China is not challenging those countries, so why would they do the dirty work for you? Unless you can give them more benefits.

However, this cannot be the reason to be blindly optimistic about the Belt and Road. "Rome was not built in a day." If we think that the situation is advantages, and expand the Belt and Road with the "Shenzhen speed", it would be a catastrophe. Skyscrapers can be built overnight, economic miracles can be created, but the Belt and Road would not become another project of wonders. We can only hope the final result is a miracle, not the speed.

This is because the Belt and Road is a process of winning people, which would be much slower than building a railway or a road. The routes are only safe when we have the support of the people. Otherwise, a "dragon" in the Go game can be cut to pieces and eaten up. The Belt and Road is especially vulnerable to this, so we have to be very cautious with each step. We should not go one way all to the end—like from Urumqi to Istanbul then to Europe—but permeate slowly with multiples routes going. Wherever we go, high speed rail ways and highways should all be built to make it well accessible, to make us less constrained by other countries. Even if one country cuts off one route, there will still be other routes. When everybody know that you would not be threatened by any one country, they will be less likely to take advantage of it.

On the contrary, if we only have one route all the way to the West, it will make us a piece of "monk flesh" coveted by the monsters. Going parallel on multiple routes give China more choices and less soft spots. More importantly, we want the countries along the Belt and Road to benefit from China's development. "All roads lead to Rome", and no one can stop us from going to Rome by blocking one. These are all necessary strategic considerations.

Fighting "Boxing" with Tai-Chi

Americans like boxing. It well reflects the love for strength—straightforward, heavy blows, and KO the opponents. Everything is clear. On the country, the Chinese people likes Tai-chi—to conquer the unyielding with yielding. I don't seek to KO you, but I'll offset your power. Tai-chi is a more delicate art than boxing, and the Belt and Road is exactly like it.

The rise of China has made us a new challenger in the eyes of the US. In order to maintain its global leadership, the US can only think of containing the challenger. It did the same to USSR during the Cold War, to Japan in the 1980s, and to the EU after Euro emerged. It wants to do the same to China. So, what would be China's response? It's to be "thrown off the bus" or to actively "get off the bus". But either way, the economy of China and needs to be detached from that of the US, which is no easy task. China has been able to become a manufacturing power today, thanks to the US market; while the US enjoyed GDP in the last 20 years doubling the peak of the 200 years before, not because of innovation, but cheap Chinese products and low-interest Chinese capital. It's a mutual dependence. Why the United States continue to issue bonds? Because they have no money. Yet the US fully understands that if they continue to print money relentlessly, the U.S. dollar will continue to devalue. They have another approach: keep the surplus in the capital account. Issue treasury bonds to borrow from abroad, building a global capital circulatory system. China is the most important link in the system. As the biggest bond holder of the US, China's position in US treasury bonds went as high as 2 trillion dollars. China once held foreign exchange reserves of 4 trillion dollars, and it was 3.84 billion dollars in 2014. This triggered questions from many people. Some people say that, with so much foreign exchange in China, why we don't spend it, instead of having the US enjoying our development? This reflects a shallow understanding of the reality that China and US are in an economic bundle. Why would China lend to the US? Not because we want to do charity. In 2008 when the financial crisis broke out in the United States, China was closely bound to the US market. It could not decouple from the US economy before the structural upgrade, and slump in the US purchasing power would impact or destroy Chinese manufacturers, especially the exporters of low-end products. Economic problems cannot be solved in an emotional way, but only step-by-step through strategies and tactics.

Go Beyond the "Zero-Sum Game" with a Mind for the World

With such a complex historical and contemporary context, President Xi Jinping proposed the strategic vision of the Belt and Road.

This is grand strategic thoughts with global vision. It is not based on the selfish consideration of one country, but with a mind for the world of "one for all and all for one". After 30 years of reform and opening up, the rapid development of China's economy has make it an indispensable player in the global resources and capital markets. The proposal of the Belt and Road is to more effectively integrate global resources and capital, and to take advantage of the existing industrial and capital advantages, so that the countries along the Belt and Road can come with China to a more prosperous future of common development. We can say that the Belt and Road is a grand strategic choice benefiting the world in the same way it benefits China.

The Great Wall

At the same time, the proposal of the Belt and Road also marked the entrance of a new stage of globalization, which inevitably would collide with the old rules of globalization. How should we understand that this contradiction between the new and the old? Zbigniew Brzezinski, the famous American geopolitical strategist reiterated the "Thucydides trap" that China and the United States was about to fall into. People have been talking about this phenomenon since the ancient Greece and Rome, that all emerging countries would eventually follow into the trap of challenging the old empire. The US once again reminded China of this, only to ask China to understand is own strength and status, and not to fall into the quandary of challenging and battling against US. But this argument can be said to be a false, as at least the United States took over the hegemony from the British Empire without falling into the trap. In particular, with the post-war "Marshall Plan" which lend money to Europe to buy American goods, it eliminated excess capacity in US and helped with the restoration of Europe. A win-win was achieved. This means that not all rises of powers would come into the trap, and zero-sum game is not the only choice. Maybe Mr. Brzezinski has forgotten that, it was only because that the US is so used to be the lead and to the winner-take-all mindset, that it came to the conclusion that China and the US is playing a zero-sum game. This is a misunderstanding by the Americans.

What's the real peril between China and the US? Not the Thucydides Trap of China challenging US, but the "hard play" by the US, which might force China to respond in the same "hard" way and lead to the outcome that neither the US nor the China would like to see.

Now, China's Belt and Road strategy is about to unveil. On the long and treacherous way forward, we can expect grand strategic games between China and the US around the Belt and Road. We must be vigilant, make sober judgments, and be fully prepared for every kind of foreseeable or unforeseeable difficulties and crises.

Liang Qiao Professor, National Defence University; Air Vice Marshal, China Air Force; Graduate, Luxun School of Literature, and Chinese Department, Peking University; Bachelor of Arts, Peking University; Recipient of special government allowance since 1992; Besides literature, Mr. Liang is also an expert of military theories. A famous military writer, theorist and commentator; Member, China Writers Association.

The Belt and Road as the Lead to a New Era of Opening up

Min Tang

The Belt and Road proposed by President Xi Jinping is a breakthrough and comprehensive long-term strategy. It is a grand strategic idea trying to build the "community of shared interests" and "community of common fate" benefiting China all related countries. Bringing together the Asia-Pacific economic circle to the east and the developed European economic zone to the west, the Silk Road economic zone is "the world's longest economic corridor with the most potential". The Maritime Silk Road joins coastal cities of China and Southeast Asian countries. It is by connections and communications over the sea, cooperation mechanism of port cities and marine economic cooperation that the Maritime Silk Road of the 21st century will finally be built.

At the same time, the Belt and Road will also promote the development and opening-up of the western region of China, reduce the development gap between regions, and achieved the grand strategy of balanced development of urban and rural areas and among different regions. Therefore, Belt and Road is a brand-new strategy overseeing the reform and opening-up, a "lead" of the new era of opening-up.

Why Do We Need the Belt and Road?

Firstly, the financial crisis broke paradigm of "manufacturing in the east and consumption in the west", reorganizing the international economy. The consumption in the United States, Europe and Japan has been declining, savings hiking, and

M. Tang (✉)
Wuhan University, Wuhan, China

W. Liu (ed.), *China's Belt and Road Initiatives*,
https://doi.org/10.1007/978-981-13-0101-8_5

it would not be sustainable to consume by borrowing. Meanwhile, the "oriental manufacturing" has been much challenged by the structural change of the international economy. China's past model of massive export and large surplus also had to change. In this context, we need to explore new markets and adjust our strategy of opening-up.

Secondly, the worsening global overcapacity calls for new demand and new drive for growth. Overcapacity is not only a headache for China, but a major issue all over the world, contributing to the recent decline in oil and commodity prices. There must be new demands and new drive for growth on a global scale.

Thirdly, international trade and investment rules is restructuring, and TPP/TTIP led by the US needs a counterbalance. The United States is vigorously promoting TPP (Trans-Pacific Partnership) and TTIP (Transatlantic Trade and Investment Partnership Agreement), which, from the perspective of their targets, requires a much higher level of trade liberalization than the WTO rules on multilateral trade. These rules are based on the developmental levels of developed countries. Given the dominance of the U.S. and European markets, the current multilateral trading system may be marginalized once TPP and TIPP are established. Many developing countries, including China, are excluded from the TPP and TTIP negotiations, risking falling into geopolitical predicaments if not responding properly. Most of these countries excluded from the TPP and TTIP happen to be on the Belt and Road, giving them an opportunity to become a counterbalance.

Fourthly, even by looking domestically, China needs a new strategy of opening-up. Over the past 30 years, China has been following a way of "bringing in", i.e. bringing in the investments and technologies while exporting the products, which is greatly challenged recently. The key reason is that China's comparative advantage has changed. With rising wages, China's labor cost averages USD 4000–5000 per year, social security and other costs included. This already very high globally, especially among developing countries in Asia. Many labor-intensive industries need to be moved out. Moreover, as a large country scarce in resources, China also needs to "go out" to get resources. In fact, China has been a net investor in foreign investment since 2014. Minimal in 2002, China's foreign investment has now made China the third largest foreign investor in the world.

Consequently, a new strategy of opening-up, namely the Belt and Road, is called for based on international macroeconomic changes and domestic needs. It has gone way beyond the literal meaning of "belt" and "road" and become the grand strategy overseeing the opening-up in the future. The Belt and Road covers 65 countries and a population of 4.4 billion, 63% of the world. International trade grows by 13.9% annually and foreign investment by 6.5%, much faster than the world average. It is estimated that in the next 10 years, the export of Belt and Road countries would amount to 1/3 of the world amount, making it a real pole.

With close to 3 billion people, many countries among these 65 have per capita GDP barely over 3000 USD, much lower than China. This suggests lots of opportunities to create a massive market.

How to Do the Belt and Road Right?

The Belt and Road focuses on five areas, namely policy communication, road connectivity, unimpeded trade, monetary circulation and understanding between the people, where the connectivity of infrastructure, cooperation on energy and resources and investments on industries and facilities led by China's foreign investment will be the core. Therefore, for a very long period, the "going out" of China's investments and enterprises should ride with the grand strategy and seize the opportunity by prioritizing the Belt and Road.

On the Belt and Road, there are long-established regional, multilateral, bilateral cooperation mechanisms. I had worked at the Asian Development Bank for many years. As early as 1990, ADB began to promote a wide range of Asia-Pacific economic cooperation, such as the Greater Mekong Delta, the Central Asian Economic Cooperation Zone, etc. Much has been done in the last 20 years, the concepts of interconnection and Asia-Pacific economic cooperation have been established, and many projects are starting to take shape. Based on these, the Belt and Road will bring the existing Asia-Pacific economic cooperation to a higher level.

To fully demonstrate the potential of the Belt and Road, a series of pioneering policies and measures would be required.

Firstly, to establish an effective intergovernmental coordination mechanism along the Belt and Road. With large discrepancies in development levels and interests among numerous countries, practical intergovernmental communication mechanisms (mainly bilateral, supplemented by multilateral) should be established as soon as possible, to match the investments and projects of China with other countries. When possible, a joint planning group can be established to finish a series of paradigm projects.

High-speed rail in operation

Secondly, the interconnectivity of infrastructure needs to be a priority. China would help countries along the Belt and Road to improve their infrastructures of transportation, electricity, communication, etc., and drive the economic integration of Asia and Europe with a series of arrangements, such as connecting the transportation infrastructures of high-speed rails, railways, highways and waterways, co-operative exploitation of resources, as well as liberalization of trade and investment. Per the estimates of Asian Development Bank, Asian countries need annual infrastructure investment of 730 billion USD, including ongoing infrastructure investments on railways and roads in China, Pakistan and Bangladesh, as well as large-scale projects such as oil and gas pipelines, power lines and communication infrastructures. The "Belt" would be an important target of investment for a very long time, and China is especially good at it. It is not only because we have strong civil engineering and design capabilities, but also massive financing capacity. Most countries would welcome such infrastructure construction coming with investments. Based on our experience in Asian Development Bank and the World Bank, infrastructure investments enjoy robust returns with great spillovers and relatively low risks.

Thirdly, different from the World Bank, the Asian Development Bank or other international organizations, foreign investments of China can get into other industries in additional to infrastructures. With infrastructure built, other industries will enjoy better development, which in turn would make the best use of the infrastructure, enabling them to repay the debts and show their economic benefits. During this period of relocating labor-intensive industries, China can expatriate some industries to Belt and Road countries, and lead our foreign investments with mature industries such as nuclear power, railway, engineering, machinery and automobile.

Fourthly, a list of Belt and Road investment projects should be compiled as soon as possible. China should push forward the construction of industrial parks in developing countries. Based on existing development zones and high-tech parks, China should encourage domestic enterprises and border regions along the Belt and Road to establish industrial cooperation zones in other countries, in the form of cooperation or joint ventures. Infrastructure investments, industrial construction, resource development and relocation of labor-intensive industries should be combined to benefit all participating countries, especially in the concerned areas such as employment, export, financial income and solvency of foreign debt.

Finally, we should actively build a diversified investment and financing framework. The Belt and Road would not work without financial support. At present, the Asian Infrastructure Investment Bank and Silk Road Fund have been or is being set up, based on which China can build a more welcoming diversified investment and financing framework of infrastructure. The government shall support provincial versions of the Silk Road funds, and fund other types of fund. China shall encourage domestic banks and financial institutions to set up branches in the Belt and Road and issue long-term bonds. More funds in the private sector should be motivated to invest as public-private partnerships (PPP). Long-term loans from international financial organizations, Official Development Assistance (ODA) and "joint financing" can also be used.

Risks of the Belt and Road?

The challenges involved in the Belt and Road should also be seen. Going into multiple areas of high regional risks, vastly different development levels and complex geopolitical situations, the cooperation is limited in both breadth and depth by the lack of multilateral free trade arrangements and effective investment cooperation mechanism. The Belt and Road may also challenge the current international order and tend to lead to strategic conflict between major powers.

The relationships between China and major powers like the United States, Europe and Japan come first. These countries are reluctant to promote the Belt and Road as their interests are infringed on. It is critical to resolve potential issues arising therefrom.

In addition, countries along the Belt and Road vary vastly in economy, politics, societies and laws, which poses great risks upon Chinese investors, whether governmental or private. With political, ethnical and religious contradictions adding even more complexity, we must be fully prepared for the Belt and Road, especially the "Belt".

Besides external risks, there are risks within Chinese enterprises. It is unknown whether these enterprises, ill-prepared for "going out" and pampered on by domestic policies, can survive in harsher environments. Generally, Chinese enterprises might pay a hefty cost if going out in large scales and quick paces before getting ready.

Moreover, there are risks of national protectionism, industries, exchange rates, finance and societies. Environmental, legal and labor welfare requirements of the host countries may also be part of the risks and challenges to the Belt and Road strategy.

To mitigate the risks, we must revise the bilateral investment protection agreements based on the situation changes. To control the political risks of enterprises "going out", the government shall establish a public information platform to facilitate risk assessment of hotspot countries and timely disclosure of risks. Diplomatic means shall be used to establish good cooperative relations with relevant countries. China shall also make full use of applicable international rules and regulations to mitigate the risks. Enterprises shall comprehensively and objectively evaluate political risks within major investment deals. Public relations activities shall be actively carried out to help getting along with local communities. Overseas Investment Insurance provided by the World Bank and the regional international development banks can be used to transfer the risks.

For the exchange rate and financial risk of Chinese enterprises "going out", export credit shall be further supported to enable more financing measures. China shall encourage banks and financial institutions to go out and provide services in areas heavily invested by enterprises. Enterprises shall reinforce management on exchange rates and financial risks, and properly control the risks with hedging tools.

Finally, the enterprises "going out" shall enhance their awareness of environment and social responsibilities, including the evaluation of environmental impacts, protection mechanism of agreements, ecological compensation, etc. We encourage enterprises to protect the environment, respects religions, faiths and customs of host countries and protect the legal rights of labors to achieve "win-win". Investment projects must acquire local environmental licenses, perform assessment on environmental impacts, and fulfill obligations of proper emission and environmental emergency management. We also suggest the enterprises to study and learn from environmental principals, standards and practices adopted by international organizations and multilateral financial institutions.

Min Tang Bachelor in Mathematics, Wuhan University; Master of International Finance and Trade, Ph.D. in International Finance and Trade and Econometrics, University of Illinois; Deputy Secretary General of China Development Research Foundation. Counselor of the State Council.

Westward Opening: A Perspective of the Silk Road Economic Belt

Wenbin Xia

The New Silk Road in New Context

The Silk Road is a major pioneering in the process of human civilization. Since Zhang Qian of the Western Han Dynasty made diplomatic visits to the Western Regions, the cross-cultural exchange had been started, thus propelling the economic and cultural exchanges and interaction between the East and the West.

Desert scenery

W. Xia (✉)
Shihezi University, Shihezi, China

W. Liu (ed.), *China's Belt and Road Initiatives*,
https://doi.org/10.1007/978-981-13-0101-8_6

For thousands of years, why would generations of people trek over mountains, rush through the desolate wilderness of Gobi, yet never cease despite unimaginable difficulties? Because the Silk Road is a road of economic development by enabling trading, a road of mind fusion with cultural exchange, and a road of adventure with exploration of natural mysteries.

Under the new historical conditions, the strategic thinking of the Belt and Road is to revitalize this road which has benefited man over thousands of years. It would be impossible to simply rebuild the inland Silk Road by Zhang Qian—the new Silk Road economic zone would be built based on the new mode of economic and cultural development, as well as the new context and trends of the world.

As the largest developing country in the world, China has created the renowned economic miracle since the reform and opening-up, adding new vigor and energy to the peace and development of the world.

Since the beginning of the 21st century, as a result of the financial crisis, the subsequent oscillation effect has kept the world economy from completely getting out of the economic downturn. After more than 30 years of rapid development, the economy and society of China have witnessed some structural challenges. Where will the world go? How can China go better and faster under the new economic norm? Precautions must be taken. The proposal of Silk Road economic belt is a major effort to influence and optimize the world economic and social development, and will bring new power to high-speed train of China's modernization.

Under the background of profound changes in international political and economic order, new trends of economic globalization and important transformation of domestic development mode, westward opening-up has become an inevitable choice for China to further integrate and participate in international economic cooperation. As Prime Minister Li Keqiang pointed out: "Westward opening-up is a major initiative of all-round opening of China." If we consider the westward opening-up in the context of steering strategy of the Silk Road economic belt, we will be able to further understand its significance, target and focus, as well as the connotation and elements of the "Belt".

The primary object and frontier of westward opening-up of Northwest China or even the whole country is Central Asia. Historically or currently, Central Asian countries hold an important strategic position in the game of powers. From the geopolitical perspective, Central Asian countries are in the center of the Eurasian continent. As early as in the 19th century, Halford Mackinder, the famous British geopolitician made a prophecy: "Who rules East Europe commands the Heartland; who rules the Heartland commands the World-Island; who rules the World-Island commands the world." This is the famous "Heartland Theory". In the ancient Silk Road, Central Asian countries brought together the artery from the East to the West", and after their independence, the modern Central Asian countries have been re-positioned as the "Eurasian bridge", the ideal economic and trade channel to drive growth with resources. Therefore, the revival of this ancient Silk Road is increasingly emphasized by Eurasian countries. Today, with overall political stability, good momentum of economic development and abundant resources, the

Central Asian countries have become a massive market of great potential, creating opportunities of strategic cooperation for China's westward opening-up.

China and Central Asian countries have a long history of peace, which is the most important historical basis for building the Silk Road Economic Belt. In 1941, the toughest time during the anti-Japanese war, Xian Xinghai, a famous musician of China, travelled to Almaty, Kazakhstan. While he was poor and sick, Bakhitzhan Baykadamov, a Kazakh composer, helped him recover and talked to him about music and culture. It was during this period that Xian wrote the famous music of national liberation "Man Jiang Hong", and wrote a symphonic poem on Amangeld, a national hero of Kazakhstan. This is a witness of the profound friendship between the Chinese and Kazakh peoples.

Vast in area and abundant in resources, countries in Central Asia, South Asia and West Asia have great potential of economic and trade cooperation. The vast land is ideal for agriculture and husbandry, making them important producers of grain, cotton and animal products. The five Central Asian countries are especially rich in energy and mineral resources, including oil, natural gas, as well as nonferrous and ferrous metals. Kazakhstan has proven 223 oil and gas fields with proven reserves of 4.8–5.9 billion tons, accounting for 3.3% of the world's total reserves. According to Kazakh experts, before 2025, Kazakhstan's proven reserves of crude oil will account for 5.5% of the word. Its total potential crude oil reserves are estimated to be between 12 and 17 billion tons, of which more than 60% is concentrated in the Caspian Sea. The preliminary proven recoverable reserves of natural gas exceed 10 trillion cubic meters, and geological reserves is between 25 and 30 trillion cubic meters.

With the accelerated pace of reform and opening up, China's rapid economic development has made it the world's second largest energy consumer, unable to meet the demand with its own reserves. Without energy, China's modernization will lose its steam, therefore, energy security is part of economic security. At present, China mainly depends of the Middle East for oil supply, and the Malacca Strait is the only path. Not only is the traffic congested and slow, but also vulnerable to external forces and uncertainties. The fact that China relies mainly on foreign cargo doesn't help. These unfavorable factors will have a serious impact on China's energy security. On the contrary, Central Asia, rich in energy reserves, is known as the "second Middle East" and one of the three major oil producers besides the Middle East and Russia. Central Asia is also at the early stage of large-scale exploitation with vast potential. These resources have laid a solid foundation for the rejuvenation of these countries, but also serve as stable "gas stations" on the Silk Road Economic Belt. Recently, China expanded in-depth cooperation with Russia and the Central Asia, making the energy sources more balanced and diversified. Build a safe and efficient strategic resource channel is crucial to the sustainable development of China and the world, and essential for lasting benefits of the Silk Road Economic Belt.

The market potential and soaring development of the neighboring countries to the west of China provide strong support for China's westward opening-up. Many of you might have been to Khorgas—I went there with high expectation a couple of years ago, yet found it was not as popular as I expected. But all had changed when I went there again with the team in 2014. I invited people doing border trading to our conference, and they mentioned their businesses were booming thanks to the encouraging policies and developing cooperation with Central Asia. At present, Horgos has great prospect, with an annual import and export of 2 million tons and people cross the border for 3 million times. One of the important reasons behind the boom is that Kazakhstan has also made great achievements. Yelbolat Dosayev, Kazakhstan's Minister of National Economy, announced on September 15, 2014, that Kazakhstan's GDP would reach $400 billion by 2019, with real growth rate of 5–6.8%. The per capita GDP would also grow from $13,900 in 2015 to $21,400 in 2019. President Gurbanguly Berdimuhamedow of Turkmenistan also announced on the government work conference on September 11, 2014 that, the GDP of Turkmenistan had grown by 10.3% since the beginning of the year. The growth rate of Central Asia has been between 5 and 10% in the recent years, while the average GDP growth in the world is less than 3%. This astounding growth in the Central Asia is similar to China—should they unite, they might have the power to influence and change the development of the world.

With rapid economic development and continuous improvement of people's living standards, the Central Asia is showing its potential as a huge consumer market. I went for 3 business trips in Kazakhstan and other countries, and I could feel the changes each time, with local people becoming richer and more confident, and they admire the development of China, with great expectation of Chinese goods and tourism in China. There are 350 foreign students in Shihezi University, most of which are from Central, South or West Asia. I have talked to them for many times on why they chose China and Xinjiang, and the answer is almost unanimous—they want to work in China and trade over the border. That would be another proof that the westward opening-up is having unprecedented opportunities.

"Two Overall Situations" and Westward Opening

As early as 1987, Deng Xiaoping pointed out that "coastal areas shall speed up the opening-up, so that this large area with 200 million people develop first and lead the development of the inland area. This is critical for the overall situation, and the inland areas shall cooperate with it. In turn, at certain stages, the coastal areas shall contribute more to the inland to help them development, which is another big situation. At that time, the coastal areas shall work with that overall situation too."

The report of the 18th National Congress of the Communist Party of China pointed out that China should "encourage coastal, inland and border areas to draw on each other's strengths in opening up; develop open areas that take the lead in global economic cooperation and competition; and form leading areas of opening up that drive regional development." That is to say, if only the coastal areas of the southeast open up, while the northwest region is always closed and backward, the progress of the development would not be able to benefit China as a whole.

We are moving towards the goal of building a well-off society in an all-round way, but for China, such a big country with unbalanced development, A well-off society for the whole China cannot be represented by the developed metropolitans, but only by the remote areas. In May 2015, I participated in the signing ceremony of the Collaborative Innovation Center of the Chinese Path and Chinese Marxism, led by Peking University and joined by top institutions such as the Central Party School, Jilin University, "Qiushi" magazine, etc., as well as universities in the Guangxi Zhuang Autonomous Region, Xinjiang Uygur Autonomous Region and other autonomous regions. A consensus was reached at the meeting: the Chinese path needs to go wider and farther—Remote areas including Xinjiang has to be modernized to truly realize the great rejuvenation of the Chinese nation.

In the "Vision and action" of the Belt and Road, the core of the 21st Century Maritime Silk Route Economic Belt lies in Fujian, and that of the Silk Road Economic Zone is Xinjiang. Xinjiang would be the key window to the westward opening-up, and become the hub for transport, trade, logistics and education. There is no doubt that, with the tide of westward opening-up in the Silk Road Economic Belt, the ancient and beautiful Xinjiang is facing great opportunities.

Geographically, Xinjiang, close to Central Asia, is the hub of transportation of the Silk Road Economic Zone. I had a lot of personal experiences—it takes 3.5–4 h to go from Urumqi to Beijing, but only 1.5 h to go to Kazakhstan. Many businessmen can make a day trip from Urumqi to Central Asian countries.

From a cost accounting perspective, land transport to Europe saves about 1/3 of the time compared to maritime transport. With the construction of the Silk Road Economic Zone, Xinjiang is building five trade and logistics centers of Yining, Horgos, Kuitun, Kashgar and Urumqi as the new engine of economic growth.

With its rich historical and cultural heritage, Xinjiang is a melting pot of a variety of cultures with multiple coexisting religions. In his dialogue with Daisaku Ikeda, Arnold Toynbee, the British historian, said that if he could live again, he would like to live in Xinjiang of the 1st century with Buddhism already there. Ji Xianlin, the renowned scholar of Peking University, said that for the great ancient Greek, Muslim Arab, Indian and Chinese civilizations, Xinjiang is the only intersection of them all. I had multiple exchanges with Mr. Ji Xianlin, and with his knowledge, I would see his comment to be well-justified. Surely, the task now is to lead this ancient culture with modern and advanced culture.

Westward Opening-Up: Goals and Tasks

For China, while enhancing the opening-up on the southeastern coast, it shall realize the complicated situation and unstable factors, which brings forward the significance of westward opening-up. The two directions of opening-up is just like the two sets of wheels on both sides of the car, or the two wings of a bird. The opening-up strategy needs to be implemented in multiple facets and directions to explore more opportunities.

Mural of Dun Huang

Another goal of westward opening-up is to support the optimization of international socioeconomic order. The rise of China as a great power is gradually improving the order, especially by the establishment of new consortiums, new national strategic cooperation platforms, such as the BRIC countries. The new platform is not to challenge the existing order, but to enrich and enhance it. It's proven that the hegemony by the US and a few developed countries need to be changed. Since President Xi took the office, he has been strengthening the status and role of China in the world peace process, and has repeatedly emphasized that we cannot fall into the Thucydides Trap, which was introduced by Thucydides, a famous historian of the ancient Greece, saying that a rising power would cause fear in an established power which escalates towards war. Some people said that China would inevitably collide with developed countries, especially the United States, on its path of development. The Silk Road Economic Zone was proposed specifically to avoid falling the trap. The peaceful rise of China will be a new auspice for the world. With its culture emphasizing harmony instead of uniformity, China would make unique contribution to the world peace. The essence of the Silk Road is peace and cooperation, from Zhang Qian of the Western Han Dynasty, Ban Chao of the Eastern Han Dynasty, Xuan Zang of the Tang Dynasty to Chen Cheng of the Ming Dynasty. Their trips to the Western Regions were to promote peace and cooperation, so is the Silk Road Economic Zone of today.

Westward Opening-Up: New Initiatives Required

In "Vision and proposed actions outlined on jointly building Silk Road Economic Belt and 21st-Century Maritime Silk Road", China has proposed five major goals of policy coordination, facilities connectivity, unimpeded trade, financial integration and people-to-people bonds. This is the guideline for the westward opening-up and also the key to success.

The improvement of policy system and laws and regulations is the fundamental guarantee. Domestically, we need to enhance the top-level design with in-depth study of the new situations and issues in westward opening-up, adjust unsuitable regulations and actively promote innovation on the system. As for the international exchange, we need to enhance the exchange on laws and policies with Central Asian countries, to facilitate the communication, improve the efficiency and simplify to be practical. This would eliminate the bottlenecks in laws and policies, and bring innovation in both China and other countries. When we discussed with entrepreneurs at the Xinjiang Border Trade Bureau, the greatest hindrance had been the lack of knowledge to the laws and regulations in the Central Asia, South Asia and West Asia. In 2014, we invited experts of law and border trade to the Central Asia to exchange views on reducing blockades on policies and laws for the border trade. The Border Trade Research Center of China was jointly set up by Shihezi University and Foreign Economic and Trade University jointly, focusing on resolving the bottlenecks of border trade in Xinjiang.

Infrastructure shall lead the way to opening-up. The construction of railway between China and Uzbekistan and Pakistan shall be accelerated, and transportation projects around the Altai Mountains shall be planned. Although the Central Asian countries is enjoying amazing economic growth and rapid improvement of living standards, their infrastructure and economy still lags behind China. The China-Pakistan Economic Corridor, extending from Kashgar to Gwadar with a total length of 3000 km, would be connecting the Silk Road Economic Zone to the north and the 21st century Maritime Silk Road to the south. This transportation hub would be a corridor including roads, railways, oil and gas pipelines and cables, and an important strategic component of the Belt and Road.

At present, China should take full advantage of infrastructure construction and actively expand infrastructure projects to the west. At the APEC meeting in 2010, China proposed the establishment of Asian Infrastructure Investment Bank, which will undoubtedly promote the infrastructure construction in surrounding areas of China.

We shall consolidate and develop neighborly and friendly relations with Central Asian, South Asian and Western Asian countries and enhance political mutual trust. As a result of various political forces, the international political situation is becoming increasingly complex. Nevertheless, China holds its determination and confidence in the path of peaceful development, which is also the mainstream of the world. To this end, we need to seek common ground while reserving differences, and adhere to political mutual trust. We shall strengthen and expand high-level visits with these countries, gradually establish and improve the political mechanism of communication and consultation, and strengthen strategic cooperation on the basis of mutual trust. Ma Junwu, National Moral Model and one of my friends, holds a post on the China-Kazakhstan border with his wife for many years. "Do only one thing all my life, which is to be a guard of my motherland.", said he when President Xi came to meet him in Xinjiang in 2014. His post is about 100 m away from Kazakhstan, and the couple guarded the boundary marker for 20 years, despite the loneliness. This is a marker of security, a marker of hope and a marker of spirit. The border posts in northwest China are symbols of friendliness to our neighbors.

We shall strengthen the security cooperation with Central Asian countries under the framework of the Shanghai Cooperation Organization, work together to combat terrorism, explore the response mechanism to non-traditional security issues, and emphasize the crackdown of the Three Evils of terrorism, separatism and religious extremism. Shanghai Cooperation Organization was established in 2001 and holds a summit every year with a focus on security and development. As the security situation in Western China becomes more challenging, we need to rely on platforms like Shanghai Cooperation Organization, and build a secure and stable zone by establishing a cooperation mechanism of unified goals and coordinated actions, an emergency mechanism of fast and effective response, and an information sharing mechanism of comprehensive information and easy operation.

We want to expand public diplomacy and promote harmonious environments. Northwestern China has a natural ethnic and cultural tie with surrounding countries, which would benefit economic and trade exchanges as well as the development of harmonious ethnic relations. We shall enhance political inclusion, understanding and mutual trust of people among different countries. When my colleagues and I visited Kazakhstan in 2014, the officials in the Chinese Embassy invited us for a visit and told us that was an unprecedented political and diplomatic "honeymoon" between the two countries. However, to continue, consolidate and develop this relationship, it wouldn't be enough to just rely on the embassy and the top leadership. We also need scholars like you to act as cultural messengers to create public and civil diplomacy and make the bond of friendship and solidarity stronger. This also reminds us to promote the westward opening-up with new approaches.

Wenbin Xia Deputy Director of CCP Committee of Shihezi University, Director of the Collaborative Innovation Center of Central Asian Civilization and Westward Opening, Professor of the Research Center of Socialism Theory with Chinese Characteristics, Vice-president level cadre of Peking University, researcher and doctoral advisor at the Collaborative Innovation Center of Chinese Path and Sinicized Marxism. "Philosophy and Social Sciences Talent of the New Century" in Beijing ("Hundred Talents Program"), and Vice President of the National Society for Theoretical Studies of Socialism with Chinese Characteristics.

How the Belt and Road Initiative Expands the Diplomatic Arena for China

Peisheng Yao

The Belt and Road Initiative Is a Once-in-a-Century Opportunity

In March 2015, while speaking at Bo'ao Forum for Asia, President Xi Jinping said, the Belt and Road initiative is not meant as rhetoric. It represents real work that could be seen and felt to bring real benefits to countries in the region. The Belt and Road Initiative focuses on the construction of transportation infrastructure, and, according to the actual demand of countries along the routes, connects China's development strategy to that of our Asian and European neighbors. A network of mutual interests is being woven, which will generate benefits for neighboring countries. We will also benefit from the collective development of those countries as well, and so, in a nutshell, collective development ensures common interests.

Some people in the foreign countries have intentionally twisted China's proposal, calling it "expansionism", "a threat" or "good old colonialism". The truth is that they don't want to see a powerful China, fearing that China will unite the developing world and seek to overthrow existing international political and economic order. They also fear that developing countries will have a bigger say in international affairs. That will make a few believers of "power-politics" very uncomfortable. Of course they'd hate to see that happen. That is why they keep causing trouble and trying to undermine China's influence in areas such as politics, the economy and the military. Since the 1990s, some western countries have been manipulating the United Nations Human Rights Council, making excuses concerning China's "human rights issues" to push anti-China proposals for ten consecutive times. Yet not once have they succeeded, the primary reason being that China has had an ever-growing global influence since the economic reform. Most developing countries have chosen to stand alongside China and have supported us.

P. Yao (✉)
Beijing Foreign Studies University, Beijing, China

W. Liu (ed.), *China's Belt and Road Initiatives*,
https://doi.org/10.1007/978-981-13-0101-8_7

Therefore, as China earn a bigger voice in the international arena, we shall never forget their precious support. Especially that of our African allies, who are our true friends.

Those who compare a fast-developing China to western colonialism claim that China will follow the same unfortunate pattern. But the fact is, China knows more about lessons in history than they do. Many Western powers chose a bloody path of selfishness in order to reach prosperity, and China was often the biggest victim. What effect did the Opium War have on China? It left China war-torn, with territories ceded. People paid indemnities, and suffered tears, starvation and death. Any Chinese person with a heart will not, and shall not forget this most humiliating part of our history. That is why, when some countries try to suppress China with human rights issues, our UN ambassador will respond with a resolute and decisive answer: "Take a look at your own history, and think about whether you have the right to criticize China."

As a matter of fact, China has been the most kind and gentle of all the major powers in the world. We never forget the favors that we have received and always give twice as much in return. This is a virtue deeply rooted in Chinese traditions and culture. Here, I find it necessary to mention the Silk Road. Although it was explored and built by both the east and west, it is a legacy of harmony, inclusiveness, equality and goodwill—largely constituted by the Chinese people. This was also the case of the Maritime Silk Road. When Zheng He led his fleet on their great expedition, it was not only to exchange products and establish connections, but also to make friends. Even though China had sufficient resources and the power, it was not interested in carrying out military invasions, taking cities and territories, or colonializing other countries.

The far-reaching tea route, which starts in the Wuyi Mountains in Fujian Province, and stretches all the way through Mongolia and Russia to Europe, is another example of China showing harmony, openness and inclusiveness. Utilizing the ancient tea route for further cooperation with Mongolia and Russia was first proposed by President Xi Jinping during his first state visit to Russia in March, 2013. He actually mentioned the idea of "one route: one belt and one road" in that same speech, which was the historic foundation of the Belt and Road initiative.

In reality, it was proposed out of the need for faster development and further cooperation between China and its neighboring countries. About two decades ago, newly-independent countries in central Asia began discussing reviving the Silk Road—this was soon after they had established diplomatic relations with China. In the Spring of 1994, I was with the delegation when our leaders of state visited Uzbekistan, Turkmenistan, Kyrgyzstan and Kazakhstan. Both sides spoke of the ancient Silk Road during these meetings. Of course, back then, the discussion was limited to how we could increase traffic volume through highways and railroads, and so is a far cry from the visionary concept it is today. However, now that both sides have stronger economic exchanges, they have seen the potential for cooperation on a larger scale. The Belt and Road initiative was proposed on the grounds of development and changes in macroeconomics. It didn't come out of nowhere. On March 28th 2015, the National Development and Reform Commission, the

Ministry of Foreign Affairs and the Ministry of Commerce co-issued a paper on "The Vision and Actions on Jointly Building the Silk Road Economic Belt and 21st-Century Maritime Silk Road" (to be referred henceforth as the "Vision and Actions"). This file provides an official and elaborate interpretation of President Xi's 2013 proposal, and describes its background as well as establishing its principles and framework. It introduces China's pursuits for "opening-up", and details China's existing cooperative resources and ultimate goal. There are detailed descriptions about Fujian province's geological advantages. I have heard from some comrades in Fujian that they're accelerating their planning process with aims to implement action soon.

A mosque in Kazakhstan

It needs to be pointed out that the Belt and Road initiative has been considered to benefit both domestic and international environments. Domestically, there is a focus on speeding up the development of central and western China, areas where either quantity or quality of the economy, fall far behind that of the eastern provinces. Fujian has developed rapidly, but there are inequalities across different areas within the province. There are also disparities between Fujian, and provinces in the Yangtze River Delta. The fundamental problem is that some provinces are not as adequately connected to the outside world. I believe that the key to China's future development is exploring these central and western regions. Such a strategy has been a key objective by the central government for decades.

Internationally, we should improve the scale and pace of cooperation with our neighbouring countries and those along the Belt and Road. Take central Asia for instance; this region has proven an outstanding example of how bilateral economic cooperation can be fruitful and promising. Though both sides still haven't reached what they hope to achieve, a breakthrough has definitely been made.

As a result, Vision and Actions became such a guideline of great importance. Naturally, there were some disputes over the issue of planning at first. Worries were expressed for possible risks concerning scale of the plan. I personally heard skeptical argument against its feasibility, which is understandable. For there has been a misunderstanding or even an illusion, thinking China wants to take on everything just because we proposed the initiative.

The "Vision and Actions" paper offers guidance on matters of great importance. Naturally, there were some disputes over it in the initial planning stage. Worries were expressed about possible risks concerning the scale of the plan. I personally heard some sceptical arguments against its feasibility, which is understandable. For there have been some misunderstandings or even illusions, that China wants to take on everything, just because we proposed this initiative.

Diplomacy as a Guarantee for Belt and Road Construction

The initiative will be carried out following the principles of discussion, joint development, and sharing of benefits. "Discussion" requires negotiations and communication with relevant countries or regions and organizations. That is diplomacy. In my view, all diplomatic activities centering around joint development can be considered Belt and Road Diplomacy. Though most countries along the routes support China's proposal and are willing to strengthen cooperation with China for the sake of development, which is the foundation of the implementation of the initiative, these countries bear considerable differences in basic reality, with most of them being developing countries that are distinct in economy, population and land area. These countries also are divers in history, culture and religion as much as they succumb to inequalities in geopolitics and economy. Some of them are allies and others neutral. Diplomatically, for China, these differences pose new questions that require profound investigation and tremendous input of diplomatic resources. After all, diplomacy serves to provide political and legal guarantee for the Belt and Road Initiative.

What are the political and legal guarantee? President Xi's "five-pronged approach"—policy coordination, facilities connectivity, unimpeded trade, financial integration, and people-to-people bond—covers multiple fields. For example, of finance, how are the currencies linked? Or, of transportation, how are the railroads connected? The construction of a highspeed railroad from Beijing to Moscow is under discussion. It stretches as long as seven to eight thousand kilometers but there's the problem of track gauges being different. Currently, when going through the border, the carriages of the passenger train have to be hung up for changing wheels. That won't work when it is a high speed train. That technical problem alone would take some effort to solve. And there are volumes of legal papers that need to be approved by both sides through diplomatic negotiations. Most of those negotiations will be led by the Ministry of Foreign Affairs with attendance of other departments. It can be confirmed that through joint contribution China's diplomatic theories and ideas of the new era will be enriched. This is a process of mutual promotion. Building the Belt and Road is of great importance in every sense. Two years ago we did not see the profundity of its meaning but now we have a clear vision of the future. It can be described as a booster that facilitates cross-border connection and accelerates overall development of the nation.

Speaking on diplomacy today, I reckon it's necessary to review the diplomatic history of New China. Since its founding, China has been persistent with peaceful diplomatic policies while being independent and autonomous. And because of that China has won worldwide acclaim. Though, due to the changing international situation, China has made some adjustment to its diplomatic policies, the principles of independence and autonomy have remained unchanged. Even during the early days when the then newly-founded China relied largely on the Soviet Union, it had never abandoned independent diplomacy. Zhou Enlai had said in numerous occasions that an alliance with the Soviet Union was necessary for resisting threat of the United States. But we should never give up independence of the party and of the state. And we should not blindly copy the Soviet experience. Strategically, we are allies, but tactically, critique is imperative. Our country was barely on its feet back then, with a corrupted society and wrecked economy left by the old regime. Even so, we didn't beg for anything from the USSR, then a superpower. Nor did we show a hint of subservience. That was a strong demonstration of the independent spirit of New China's diplomatic policies. Surely, the Soviet Union made a considerable contribution to Chinese economy during the early days of the PRC, which we have never forgot. Soon ago when president Xi Jinping visited Russia for the Commemoration of 70th Anniversary of World Anti-Fascist War, he deliberately arranged a meeting with Soviet soldiers who fought in the War of Resistance Against Japanese Aggression and some Soviet experts who came to the aid of China.

Since Deng Xiaoping launched the economic reforms in 1978, China's diplomatic policies went through some serious adjustment, mainly the establishing of "non-alignment, non-confrontation and transcending ideology" as the guideline, through which China won more friends as well as bigger international political space. Without the adjustment, it was unlikely that China could seize the chance of development in a relatively peaceful and stable international environment. In 2001, China and Russia signed the Treaty of Good-Neighborliness and Friendly Cooperation, the core spirit being the two sides shall stay friends to each other and never become enemies. I've been asked by many of my comrades why China and Russia have such a close tie now and if we should consider our own interest. I told them that for sure we should consider our own interest no matter how great a relationship we have with another country. Establishing a comprehensive strategic partnership is demanded by the development of international situation. It is also to ensure current and long-term interest for both countries. Furthermore, it's in favor of regional and international stability. The consequence of confrontation between China and the USSR was already proven in history. In May 1989, when meeting with Mikhail Gorbachev, then the Soviet president, Deng Xiaoping said: "We should buried the hatches and explore the future together." Those were the remarks of a great strategist, not just any politician. Deng's statement concluded the negative part of history between the two countries as embarking upon a promising prospect of joint-development. The non-aligned policy we have for Russia now is the result of a tough lesson on both sides' part. When meeting with Putin, Xi Jinping and the Russian president both emphasized further improving the comprehensive strategic partnership. This guideline will remain the same for China and I don't think Russia wants it to change, either. When two neighboring countries are this big, they can gain only from cooperation. Any kind of confrontation will always end in loss on both sides. In my view, after years of diligence, China, Russia and other countries in central Asia have become a community of shared interest and destiny, a fact that is marked by both sides showing similar standpoint in critical international issues and shouldering due responsibility in international affairs as well as respecting each other in bilateral exchanges, strengthening mutual trust and carrying out practical cooperation in the spirit of equality and mutual benefit.

Now history has evolved to the point where China is the second largest economy in the world. How do we work more closely together with the rest of the world and establish more connections? Should China follow its traditions or seek new breakthroughs via constant innovation? Historical experiences home and abroad have proven that be it large entities like a country or a nation, or smaller ones such as an enterprise or an organization, without innovation there can be no vitality, so to survive one can only follow others around. As we are seeing today, flaws of the US political system are becoming way too evident, with tons of domestic problems. The fact that the US still remains an energetic society and keep leading the world in economy and technology owes mainly to their will to innovate. Lee Kuan Yew once said that often in difficult times can the Americans display amazing innovative capability. I believe that has a great deal to do with their talent policy. The US has always been open to elites all over the world. Elites are individuals with independent and creative spirit especially in science. Their ideas are improbably far ahead of their counterparts and they can always break through the noise.

The CCCPC, led by General Secretary Xi Jinping, laid out new strategic plans for China's future development from a new historical standpoint based on 30 years of experience in economic reforms. To implement these plans, a series of adjustments were made on diplomatic policies. Xi Jinping's proposition of establishing international relations centered on mutual-benefit and win-win is a great contribution to China's diplomatic theory and thought. This proposition provides guidance on stabilizing relations between China and other great powers, and, promoting ties between China and countries in its neighboring region and the rest of the world, as well as on dealing with challenges of the Belt and Road.

Here, I'd like to share some of my personal experiences in diplomacy. Being a diplomat is an honorable profession envied by many. But what qualities should one possess to be able to work in diplomacy? Zhou Enlai once told Chinese diplomats that they should hold their ground and grasp the essence of policies and be proficient. I'd like to focus on the first rule. A diplomat, above all, must be loyal to your country and the people. Only by putting the interest of the nation at first would you be as devoted as possible. You should always be prepared to sacrifice everything, including your life, for your country. Since its founding, China has formed a team of firm-minded, disciplined, well-trained and proficient diplomats, a great many of whom have proved their worth with heroic deeds. There was this ambassador named Fu Huaqiang who once worked in the Chinese embassy to Chad. In 1980, a civil war broke out following a coup. As he was bravely coordinating the evacuation of embassy staff he was shot in the back and collapsed bleeding. Then he was rushed to the hospital for a quick treatment, thinking that it was merely a scratch. Yet afterwards he started to sense constant pain in the back. It was until 2003 during a routine check-up the doctor discovered, in his back, a bullet, which had been there for 23 years. I think one of his greatest virtues is that he never brags about anything or praises himself, instead he is always modest and diligent. He was promoted much later than other colleagues who joined the ministry together with him yet he never complained. That's why the powers that be decided to officially call on all party members to learn from Fu Huaqiang as a role model.

I've always heard people saying how glamorous it is to be a diplomat. This is understandable because this profession does come with the perk of traveling the world and learning incredible things. But what most people don't realize is that it also comes with much arduousness. Without the spirit of utter devotion as well as strong mind and physicality, one is unlikely to be competent. For example, the Protocol Department's job may seem simple enough at first because basically it's just making arrangement and holding banquets and receiving guests and sending them off. In fact, it is an extremely exhausting diplomatic task because of frequent high-level reciprocal visits. Event after event throughout the year, especially multilateral occasions, where everyone must stay intensely-focused the whole time. This task is not up to one side. Sometimes pre-determined schedules might be restarted from square one due to unforeseeable changes. So often would all previous efforts go to waste. To ensure there are no mistakes, they work around the clock on tight schedules. Normally in the same timeframe they would get multiple times more work done than their counterparts in other departments. They barely eat or sleep enough doing what they do. Behind every successful event such as the 2008 Olympic Games, the 2010 World Expo and the 2014 APEC Summit, the Protocol Department all made a tremendous contribution. I believe their sense of responsibility and spirit of devotion come from strong patriotic commitment for they understand how an irreversible loss can be caused by one seemingly-insignificant fault. Each drop of sweat they shed is for the elevation of China's national image. Many foreign friends and heads-of-state have praised that China's diplomatic protocol service is among the best in the world. I've visited other countries as part of a delegation and received our leaders during their visits while I was stationed overseas. So I know what this job is all about. Of course, other departments in the ministry, like Consular Affairs, have their fair share of hard work too. In recent years, in order to protect interests of Chinese expats around the world, the leadership of the Department of Consular Affairs and other divisions, or even the ministry authorities, often have to get up in late-night hours to deal with all sorts of emergencies.

Then, would embassies in foreign countries have an easier job? Absolutely not. I can tell you some stories.

Before the 1990s, to work in an embassy means you are separated with you family on a temporary basis. You could be gone for two to four years, sometimes six or even eight years. The only contact with your family is through letters. Back then there were no tools of modern communication, let alone mobile phones and computers, and a month's salary was barely enough for a few minutes of international long-distant calls. The homesickness they felt was unimaginable for people today. How were letters from home delivered to the embassy staff? It was actually a tag-along task when diplomatic couriers delivered confidential documents. And only once a month did couriers go to smaller embassies whereas for big ones in the US, France, Germany, the UK and the USSR couriers would visit weekly. By calculation, those who worked in smaller embassies would receive just twelve letters at most annually. That's why the days when the courier came were the most exciting and expected moment for people at the embassy because they brought with them precious messages from families back home. Economically, it was a difficult time for China so the budget for diplomacy was quite limited. And because of that most diplomats were stationed overseas alone, long unaccompanied by their spouse or children. As a result, their children couldn't even recognize their faces and when they grew up they usually went to live with their grandparents. These cases of estrangement, sometimes permanent, happened more times than they should.

And, in some African countries, unforgiving weather and horrible diseases took a serious toll on the embassy staff's health, threatening their lives. In some countries they were easily infected with malaria. The morbidity rate was as high as 90%. Some of them suffered from the disease for the rest of their lives and others even died on post. The now notorious Ebola virus once spread to many African nations where there were Chinese embassies. Faced with this lethal virus, the embassy staff could not retreat, instead, they had to carry on the hard work. The increasingly rampant international terrorism was another threat to the diplomats' lives. Therefore, diplomacy is a highly-dangerous occupation. As stipulated by international law, any diplomat above the retinue level is entitled to a privilege that strictly forbids any form of harm to be inflicted upon the person, also known as diplomatic immunity.

Instead of whining about hardships of diplomatic work, I'd rather tell you that most Chinese diplomats, wherever they are, take pride in enduring miseries and they're more than capable of withstanding the test of any challenges and difficulties that come their way for they keep the motherland at heart at all times. With China's national power growing, Chinese diplomats have been receiving better treatment since the 1990s. Today, conditions provided for them are basically at the same level of their foreign counterparts. Not only can their spouse stay with them, but their children can also attend a local school and their parents can visit them overseas.

The Overall Strength of a Nation Is the Guarantee for Their Diplomatic Endeavors

I've held a total of 12 receptions during my post as the ambassador and I spoke at each reception. Every year as I spoke I felt more confident and comfortable than the last time. Why? Because the guests were sincerely congratulating China's accomplishment in all sectors. Representatives from the ruling party and the opposition both praised Deng Xiaoping's strategic vision. Without Deng there would be no economic reforms and without the reforms China wouldn't have been what it is today. They saw that clearly. I'll give you a true example. I once knew this high-level foreign official who was, strictly speaking, a believer of anti-communism. So I arranged a trip to China for him and his wife. After they returned he told me in a quite emotional tone: "I would've joined the communist party too if I were born in this country." And I believe he was being honest. Only by comparison can one make judgments and his comment was the result of comparing China and his own country.

We should be proud of our country's fast development but not be conceited because on so many levels we're still a developing country. Therefore, we should have a calm view on China's achievements. Some experts at home say that China will soon become a superpower, even one that will overtake the US in ten years. This exciting prediction is understandable but for China to really catch up with the US it'd undoubtfully require several generations' unremitting efforts. The disparity is obvious not just in quantity but in quality as well. We still can't manufacture core components of many products such as the engine for large aircrafts. And it takes time to overcome these challenges. Even if we do make some substantial progress in both soft and hard power it'd still be unwise for China to call itself a superpower. During a speech in the US in the 1970s, Deng Xiaoping stated in a determined tone that China would not seek to be a superpower even when it's developed. In my opinion, a superpower, by definition, is not a country that tends to treat others equally. The problem with superpowers is that they think they are above others so they like pointing fingers and bullying, which we shouldn't do. We should stick to the principles of treating everyone equally and solving problems rationally even when we become a real powerful country. China has been advocating the establishment of new international political and economic order. That means global issues should be discussed and decided by all countries, not just one or two. For international financial institutions established after World War II, like the IMF, their reforms must be approved by the US Congress. Is that fair? I believe the Asian Infrastructure Investment Bank proposed and co-founded by China will not be a club of rich nations but a co-managed platform of investment and financing for all members and it will certainly succeed.

Peisheng Yao is a Russian major graduated from the Beijing Foreign Studies University. Between 1995 and 2005 he was a Chinese ambassador successively to Kyrgyzstan Latvia, Kazakhstan and Ukraine. He is currently, Director of the Asia–Africa Development and Exchange Society of China, Foundation for International Studies, Center for Contemporary World Studies, Public Diplomacy Association a, a researcher at the China, a guest researcher at the China, and a member of the China.

The Idea of "Transparent Ocean" Drives Future Expansion for China

Lixin Wu

Throughout Chinese history, never have we been so aware of the importance of oceans today, nor have we ever been so close to becoming a maritime power. Ocean researchers like us should have the sense of responsibility as well the confidence to take on historical tasks trusted upon us in this crucial time of China's ocean strategy development. Today, I'd like to talk about oceans on the following four subjects.

Carbon Emissions and Global Warming

The 21st century is the century of oceans, or, in a more direct way as President Xi put it, the 21st century is the Pacific century. From this point of view, oceans being part of China's overall development strategy is not only a path to the great revival of the Chinese nation but also the demand of the times. We have given oceans such names as the cradle of life, the homestead of storms, the controller of climate, the treasury vault of resources, and the artery of transportation. Above all, an ocean is a highly intricate system that comprises movements of various scales, which transport energies and matters in the ocean. Despite a veteran of ocean researches for many years, I'm still amazed by them, especially when looking from a global perspective today.

When it comes to the sustainable development of oceans, we should at first establish the problems facing humanity in the 21st century, the foremost challenge being climate change. On the 2014 APEC Summit, China vowed to peak carbon emissions by 2030 and the US pledged to reduce greenhouse gas emissions by 28% of 2005 levels. "A cooperative relationship between China and the US can be the ballast stone for global stability." Therefore, climate is such a vital issue that it often determines relationships at multiple levels such as economy and society.

L. Wu (✉)
Ocean University of China, Qingdao, China

W. Liu (ed.), *China's Belt and Road Initiatives*,
https://doi.org/10.1007/978-981-13-0101-8_8

The second biggest challenge would be with water resources. Fresh water resources on the earth have been decreasing. In the past, all major water-circulation research plans around the world seldom include oceans. That is a mistake both scientifically and administratively. As a matter of fact, 86% of global water circulation comes from vaporized sea water and 78% of rainfall ends up in oceans. From this perspective, oceans control the global circulation of water.

The third challenge is with energy sources, which usually determine relationships among countries. Shifts of an overall energy strategy of the US greatly affect its strategies concerning the Middle East. Oceans are primary sources of energy, especially sustainable clean energy. Another problem is food. Statistics show that there are totally 40–60 billion tons of biological resources in the ocean, where it is home to 180,000 species, three times more than on dry land. Those together can provide 1000 times more food for human beings than all the cultivated land on the earth and it would be a main source of proteins for humans in the future. To realize the dream of becoming a maritime power, China needs to have a profound understanding of oceans as well as advanced ocean-related technologies.

At the 2014 Conference of Academicians of the Two Academies, President Xi emphasized the issue of innovation-driven development. The speech put a heavy load on our shoulders. What kind of roles should the researchers of ocean sciences play in helping China become a maritime power?

Summarily, there are six frontier issues of oceanography in the 21st century. First, acknowledging the changes taking place in oceans, climate and water-circulation. Second, observation and prediction of oceans. That means observing phenomenona and processes of the ocean as well as predicting related events. Third, complicated offshore problems. Nearly 40% of the world's population live within a 100-km distance from coasts. Increasing human activities have seriously affected the offshore environment. Fourth, the imbalanced ecological system. Never should we sacrifice the environment to become a maritime power, otherwise it would not last. In Beijing and many places, the first thing people do every morning is checking the level of PM2.5. We can't do to the ocean what we did to land already. Not only do we need healthy oceans but we must make sure they stay that way. Fifth, studying the ocean floor. Discoveries of hydrothermal vents and cold springs have renewed our knowledge of the ocean. The ocean floor is leaking. Sixth, the oceanic crust and plate mechanics. Behind these scientific issues lie the problems of living environment, food as well as mineral resources and energy.

Science and social/economic development are complementary. As for the relationship between ocean and climate, it's safe to say that it is a subject of super science today for it includes problems in almost all academic disciplines. We are currently more concerned about carbon emissions and global warming. In fact, the ocean absorbs about 50% of carbon emitted in human activities and over 90% of surplus heat caused by increasing greenhouse gases. So the ocean is a sink for both carbon and heat. Vaporized sea water is a main source of moisture in the atmosphere and an important driving force for atmospheric circulation. Heat exchange between ocean and climate, which can regulate long-term climate changes, is one of the most crucial parts of energy flow in the earth's climate system. China is located in the East Asian monsoon region where moisture mainly comes from west Pacific, South China Sea and the Indian Ocean, so any shift of temperature in these waters will cause changes of moisture channel in East Asia.

Without thorough study of these changes in the arteries of moisture transmission, predicting climate change in China would be impossible. As for typhoons, ones that of normal intensity are not to be feared but a super typhoon is something different. In 2005, hurricane Katrina destroyed New Orleans overnight. In 2013, Typhoon Haiyan took away thousands of lives. Will super typhoons like those happen more frequently in the future? That's anybody's guess. But it is something that developing countries must understand and that calls for the study of oceans. To the Chinese, 1998 was a year that was not to be forgotten. Almost half of China was submerged by flood. Should a disaster of the same scale happen again, it would be a tremendous challenge for the country's economy. Also in 1998, while China struggled with flood, Australia was enveloped in sandstorms. One crucial cause of such polarity was the change in Equatorial Pacific sea surface temperature, also known as the El Nino Phenomenon. Will super El Nino events happen more often in the future? Unfortunately, according to recent findings, with the incessant increase in carbo dioxide levels, such super El Nino phenomenon as the ones in 1997 and 1998 will occur more frequently.

Fish shoal in the ocean

The Northwest Pacific Ocean has witnessed ocean temperature changes as well. In the mid-1970s it went through a change from cold to warm, which was known as the mid-1970s abrupt atmospheric temperature changes. And this phenomenon was first discovered not by oceanographers or meteorologists but by aquaculturists for they noticed an abrupt change in species and numbers of salmons in the north

Pacific region. In fact, it was a change that was also detected in the Atlantic and the rest of the world. To learn the frequency and duration of this abrupt change in the Northwest Pacific Ocean temperature in the future will provide important scientific guidance for management and exploration of marine fishery resources. Currently, we only have limited knowledge of the cause and duration of such abrupt changes, let alone being able to predict. If we could somehow establish an observation and prediction system for the Northwest Pacific Ocean, we could make it "transparent" thus we'll have a better grasp of changes in environment, climate and resources of this part of the ocean.

China's oceanographic study must extend to deep and open seas. International politics aside, over the past five decades, the average surface temperature on Earth increased by 0.5–0.8 °C. We felt warmer but not too obvious. Yet without deep-sea warming canceling out warming at the sea's surface, the temperature could have increased by 5–8 °C and the repercussions would be catastrophic, even causing fundamental changes in social and economic structures.

In the last decade or so, one major issue of climate change has been widely argued all over the world. According to temperature statistics, since the end of 1990s, although carbon dioxide levels kept rising, the global average temperature was not getting any higher. This was called global warming hiatus. If this situation were to last for 20–30 years or even longer, it'd help China's economic development and somewhat lower the pressure on reducing carbon emissions. So, either from the angle of China's development or the world's as a whole, research on aspects such as mechanism, duration and social/economic/environmental effect of global warming hiatus must be carried out in a timely fashion. Lately, professor Chen Xianyao of the Ocean University of China Key Laboratory of Physical Oceanography published an essay on the Science Magazine, putting forth that global warming hiatus is closely related to heat being absorbed by North Atlantic Deep Water. The article caused such a stir globally that the process was referred to as a "deep-sea bomb".

Where had all that heat gone in the deep sea? Being the primary heat sink of the earth, will North Atlantic become saturated? How will this "deep sea bomb" affect the environment of the deep layer? The movie *The Day After Tomorrow* demonstrated a series of disasters and the coming of ice age caused by global warming and halt of abyssal circulation. Unfortunately, we know almost nothing about the ocean from 2000 m below. Will it be possible for us to turn that part of the ocean "transparent" in five to ten years?

More on offshore issues. If deep and open seas are strategic problems, then offshore is a problem we must face on a daily basis. Currently, China has to deal with multiple pressures concerning offshore, the prominent one being the well-being of the ecological system, which suffers from eutrophication and severe pollution. Furthermore, the temperature offshore has been increasing distinctively in China over the past 30 years. The growth was five to ten times faster than global averages. And this rapid warming-up has brought a series of problems such as instability of fishery resources and marine gas hydrate. For example, jellyfish are known to be fond of high temperature in which they grow faster. That means in the next few decades China's offshore areas are likely to be occupied by jellyfish if the temperature keeps increasing. And that would change the ecological structure there, posing a serious threat to fishery resources. As we talk about ocean management today, I personally have expressed, on many occasions, the same worry that we might be walking in the wrong direction due to the lack of substantial foundation of oceanography.

Ocean management should be based on one premise. Whether it is a sound marine ecosystem or management of marine resources or rights, we need to make sure that any decision is tightly integrated with oceanography. We must include factors of climate, environment and resources when studying the ocean today. Exploration of offshore renewable clean energy is a crucial measure to take in tackling climate change. According to investigations and projections by the State Oceanic Administration Special Project 908, China's offshore region contains a generating capacity of approximately 15 trillion kilowatts. However, when exploring energy sources in the ocean, we must have a comprehensive understanding of status, processes and changes concerning the ocean, be it marine current energy, wave energy, thermal energy, tidal energy or even marine biofuel energy.

When speaking of sustainable development, we are actually referring to three aspects: economy, society and the earth's life-supporting system, each connected to one another. For oceans though, the three aspects would be blue economy, the ocean and human society, as well as marine life system. Today, one vital scientific issue brought up by sustainable development of the ocean is the prediction of multiscale change of marine environment against a backdrop of global change as well as climate resource effect. In this context the term multiscale means time, region and space. The timescale ranges from second, minute, hour, day all the way to year or even larger scale like century. The region scale ranges from coast to offshore, deep and open seas and polar. The space scale ranges from swift current to mesoscale, basin and planet. Those are determined by the complexity of marine system itself. To solve this critical scientific problem, we need to turn the ocean as we know it into the "Transparent Ocean".

Developmental Bottleneck of "Transparent Ocean"

The "Transparent ocean" refers to real-time or quasi-real-time acquisition and evaluation of marine environment data of different space scales as well as the study of multiscale changes and the mechanism of climate resource effect. Based on the above data and study, we should be able to predict time-space changes of marine environment, climate and resources in the foreseeable future, thus make the ocean transparent. Therefore, the "Transparent Ocean" consists of three closely-connected levels, respectively transparent status, transparent process and transparent changes. Effective ways to realize make the ocean transparent include studies that combine observation, perception and prediction. I don't think this is a new concept per se.

It's just that in previous development strategies China hasn't been able to integrate the three aspects.

The realization of "Transparent Ocean" requires studying coordination among the following aspects: observation, mechanism and prediction, development of science and technology, multi-interdisciplinary studies of marine ecosystem, oceanography and talent cultivation. In China, observation, mechanism and prediction are loosely linked due to institutional defects. Observation data lacks continuity and systematicness and is shared only in a limited fashion. Key observation instruments abroad are not accessible to Chinese scientists. Science is divorced from technology development, and scientific research and personnel training are not closely connected. Plus, establishing a system for training multi-interdisciplinary talents is imperative. All of these factors have co-constituted the bottleneck that restricts the realization of "transparent ocean."

Oceanographic observation is extremely complicated and costly, so in a certain sense it reflects a country's economic power. An oceanographic observation system normally includes an under-water movable platform, a fixed platform, a satellite observation platform as well as data integration and products. The Argo Program, in particular, is a revolutionary progress in the history of global oceanographic observation. With the program, for the first time ever, scientists while sitting in their office can be aware of any change that occurs in the world's oceans to a depth of 2000 m. The Argo Program planned to deploy about 4000 floating beacons worldwide and that objective is almost completely achieved now. Most floats are deployed by developed countries, among which only about 100 are from China. This is utterly unbecoming for a country of China's standing.

Currently the Argo floats can only observe to a depth of 2000 m. Now the international community is brewing the second round of the Argo Program, which will extend the range of observation beyond current floating depth. The US, Japan and France are developing deep sea Argo floats in order to gain the upper hand on the upcoming competition in deep sea exploration. Being a responsible global power, how should China move in the latest deep-sea Argo observation plan? It is not just a challenge but a rare opportunity for China to make a great leap forward and proceed to the vantage point. By turning the ocean below the depth of 2000 m "transparent", we will unveil its mystery and have a better understanding and grasp of deep-sea environment, biosphere and resources.

For a considerable period of time, the Pacific Northwest has been a critical factor concerning China's core interests in the ocean. Yet after so many years the marine observation system in the region has not been established. Today, China's space industry is developing rapidly. We fly higher and farther than we ever could. As an oceanographer, I do hope we can dive deeper and farther as well. I have a dream that China will establish, in the West Pacific, the South China Sea and the Indian Ocean, an observation system that can support our national maritime strategy. As a responsible global power, China should make its due contribution to sustainable development of the world's oceans. The Tropical Pacific Moored Buoy Array can measure real-time temperature and wind speed of sea surface/sub-surface and then transmits the data to land through satellite. This provides vitally important information for the forecast of ENSO, or El Nino Southern Oscillation.

However, due to the lack of maintenance funds, nearly half of the buoys are inoperative. Buoys in the West Pacific Warm Pool were originally deployed and maintained by Japan but now are facing financial trouble as well. China should think and plan carefully as to what we can do to help. At the present time, marine observation doesn't simply rely on traditional instruments but requires a revolution of new technology. In the first half of 2014, when addressing at the Conference of Academicians of the Two Academies, President Xi mentioned robot technology and big data as he talked about innovation-driven development. Both technologies are especially important to the realization of "Transparent Ocean".

Predictions are of the same importance as observations. After MH 370 went missing, the whole world was hoping to find the wreckage. The sea area where the plane went down can be pinpointed using particle tracking technology, which was employed for the search of Air France Flight 447. The fastest computer in the 20th century was the Earth Simulator developed by Japan. Manufacturing of the device was advised by Japanese meteorologists. Then there was the Living Earth Simulator of EU and Blue Gene of the US. Currently, the fastest computer in the world is Tianhe-2 in Tianjin, China. One main function of these super computers is simulating and predicting changes in the earth's climate system. Simulations and predictions of the ocean, which is an important part of that system, need those super computers. The realization of "Transparent Ocean" will not be possible without the support of VLS computers.

Some might ask what exactly is the "Transparent Ocean"? Does it mean polluted sea areas turning transparent through restoration? That's not correct. The "Transparent Ocean" figures out the status, processes and changes of the ocean using complicated observations, data and prediction systems. The ocean is so vast, deep and complicated that we must have a well-thought-out plan to make it transparent.

Implementation of the "Transparent Ocean" Plan

The "Transparent Ocean" is a large scientific plan that needs to be carried out in a gradual and organized way, advancing from the transparent shelf sea and the South China sea to the transparent West Pacific, the Indian Ocean as well as the Antarctic Ocean and the two poles. A few days ago, when visiting the newly-founded School of Marine Science and Technology of Tianjin University, I spoke of the future development and asked them if they could integrate research teams in Tianjin to realize the "Transparent Ocean". The development of the Bohai Economic Rim has laid tremendous pressure on the region's ecosystem, pushing its carrying capacity to the limit. "Transparent Bohai" is the key to transformation of environment and resources against the double threat of human activities and climate change, as well as an important guarantee for maintaining sustainable growth in the region.

In fact, the realization of "transparent shelf sea" in China will not only benefit Bohai but also provide critical support for sustainable exploration and management of China's eastern shelf seas. But up to this day the long-term shelf sea observation and prediction systems are still far from complete. This incompleteness is mainly reflected in interdisciplinary and long-term consecutive observation. For example, the Yellow Sea Cold Water Mass is the most distinct phenomenon of China's eastern shelf sea. Shifting in its temperature and spatial location greatly affects fishery resources. But we haven't been able to establish a multidisciplinary long-term consecutive observation system for this important phenomenon. And this has restricted the level of prediction of its alteration and the ensuing change in fishery resources.

One important task for marine observation is to provide for prediction. As a facility of strategic importance, not only will Ocean University of China realize the "Transparent Ocean", but it will broaden its horizons and work hard to further full implementation of the plan of "the transparent West Pacific, the South China Sea and the Indian Ocean". To that end we have taken full advantage of Qingdao's position as a key marine research base in China and co-founded Qingdao Collaborative Innovation Center of Marine Science and Technology along with Shanghai Jiao Tong University, Institute of Oceanography of CAS, The First Institute of Oceanography of SOA and Peking University. In order to accomplish the mission of realizing the "transparent West Pacific, the South China Sea and the Indian Ocean", we have drawn up a four-point strategic plan as follows.

One, technological breakthroughs. With current technology deemed incapable of supporting China's maritime ambitions, it is imperative to research and develop key devices and technologies for the deep-sea observation system, especially underwater floating platforms. We need to have independently-developed products and improve out observation ability so as to break the technical barrier.

Two, expanding the observation network. Temporal-spatial resolution of the network should be improved. Single observation should be expanded to multifactor comprehensive observations. A wholesome, real-time and multidisciplinary observation network is to be formed.

Three, theoretical breakthroughs. Carry out coordinated studies of environment, climate and resources in the West Pacific, the South China Sea and the Indian Ocean. Strive for significant original innovation in such aspects as mechanism of multi-scale change in marine environment and climate resource effect.

Four, building prediction systems. Gradually and in an organized fashion, we will set up a climate prediction system for West Pacific, South China Sea and the Indian Ocean, as well as prediction systems designed for specific demands of China. In summary, it will be a multitiered, multidisciplinary and multipurpose prediction system.

With strong support from China's maritime power strategy and through incessant effort to keep abreast of international standards and cutting-edge oceanography, the implementation of the "Transparent Ocean" project has made a series of periodic progresses. Driven by the demand of observation network construction, a number of critical technologies are evolving rapidly and the mechanism for sound development of marine science and technology has been formed. A preliminary observation system for the West Pacific, the South China Sea and the East Indian Ocean has been established. Our observation ability is improving fast and a breakthrough has been made in the study of changes and transparency of the ocean. In the past three years, a number of original achievements concerning the interaction between global warming and the West Pacific, the South China Sea and the Indian Ocean have been made. Global influence of our work had been increasing. Coordinated development of the pattern system laid a solid foundation for the establishment of prediction systems.

Launching international cooperative plans has brought a great many of scientists to the center stage of the world. In October 2014, the Center held the Global Ocean Summit at the 90th anniversary of OUC and for the first time gathered all leaders of top ocean facilities in the world. The summit discussed developments of the ocean and accelerated world-wide implementation and coordinated development of existing marine observation structure, thus provided support for organic integration of factors in current marine observation system and mechanism innovation in the course of implementation. Meanwhile, the summit demonstrated numerous new fields of research in current development of oceanography that are significant to serving the society. By including these new observations and research into existing observation system, not only will we have access to more data and enhance development of the ocean-climate response model, but also increase support for and investment in the existing observation system.

The summit calls upon global researchers to explore and collaborate on sustainable utilization of the world's marine resources. By establishing a sustainable and comprehensive global marine observation system, we can examine, predict and respond to long-term development of the ocean and maintain sustainable growth of the earth's ecosystem. The summit brought about the signing of the Future Marine Qingdao Consensus, which centers on Premiere Li Keqiang's speech on scientific development of oceanography in China. Through cooperation among international research institutions, a peaceful, harmonious and cooperative partnership is to be established, pushing the "Transparent Ocean" project to a deeper and further level.

On the Future

Under systematical planning and guidance, the implementation of the "Transparent Ocean" project has made productive progress. As for goals and next step of the "Transparent Ocean" project, I believe there are four aspects to be considered. One, forming a batch of technologies for marine observation and detection as well as key prediction devices so as to boost the development of China's marine equipment industry. Two, establishing an observation and prediction system for the West Pacific, the South China Sea and the Indian Ocean, thus forming the "Transparent West Pacific, South China Sea and the Indian Ocean" as well as the capability to support oceanographic studies and critical national maritime strategic demands.

Three, establishing the fundamental theory for multiscale energy and matter transmission and exchange in the West Pacific, the South China Sea and the Indian Ocean, as well as forming a new world-class cutting-edge interdisciplinary system of marine environment, resources and climate, so as to guide the development of China's deep sea technologies. Four, establishing three to five world-class deep-sea research and innovation teams and establishing a new model for cultivating interdisciplinary talents of deep-sea studies, forming a new model for recruiting and cultivating talents that meet the demands of innovation in oceanography and technology collaboration, and becoming one of the most active platforms for international oceanic exchanges and cooperation.

Huge waves of the ocean

Traditionally, Oceanography includes four subjects, namely, physical oceanography, marine geology, biological oceanography and chemical oceanography. However, present development of oceanography not only requires profound intersection of the four subjects, but also needs support from technical subjects such as molecular science, robotics, communication, nano materials and satellites, as well as integrating with relevant subjects such as atmosphere, math, information, life science and management. Only with effective interdisciplinary studies can we carry out oceanographic research and meet the demand of China's strategic development, as well as solving problems for mankind. Everyone can find a place in the development of oceanography and China's rise as a maritime power, both require our collective effort. The "Transparent Ocean" project was proposed in response to China's maritime strategy and is a necessary step towards sustainable development. I do hope my interpretation would provoke more thoughts and challenges, and provide a clearer definition of "Transparent Ocean" in concept and theory as well as in practice.

Lixin Wu is a physical oceanographer, a professor at Ocean University of China and an academicia at the Chinese Academy of Sciences. He obtained his Bachelor of Engineering in Mechanics from Tsinghua University. He holds Master and Doctoral degrees of Mechanics from Peking University. He has eleven years of scientific research experience in the US. And currently, he is a top-level professor of the Peak-building Project at Ocean University of China (since 2005); a recipient, the National Science Fund for Distinguished Young Scholars; a Special-term Professor, Changjiang Scholars Program of the Ministry of Education; an academic Leader of Innovative Researches, National Natural Science Foundation of China, Academic Leader of Innovative Researches in Key Sectors, the Ministry of Science and Technology.

Guided by Shipping and Guarded by Law—Legal Thoughts on Building the Modern Maritime Sild Road

Wei Zhao

In this new era of political multi-polarization, economic globalization, and complicated maritime conflicts, China, by putting forth the strategic idea of rebuilding the 21st century Maritime Silk Road, has demonstrated grand ambition as a maritime power that respects history and pursues peace as well as a will to benefit the people in countries along the route by developing together. It has become so evident to us that the core motivation of building the Maritime Silk Road lies in setting up a platform of maritime economy and trade for countries in East, South and West Asia, East and North Africa and Europe, as well as promoting commercial and cultural exchanges among those countries. The ultimate goal, under the principles such as "equality and cooperation, mutual benefit and win-win, openness and inclusiveness, harmony and friendliness", is to realize technological innovation, cultural convergence and trade exchange, bringing tangible benefit and spiritual joy to the people. However, this great strategic goal cannot be accomplished without improvement of China's maritime transport capability as well as protection of the legal system.

The Very First Chapter of Building the 21st Century Maritime Silk Road Is Opening More Sea Routes and Port Construction

Although merchandise and services are the focus and theme of trade exchanges and economic cooperation, the top priority when building the 21st century Maritime Silk Road is opening more sea routes and port construction because without shipping there are no exchanges and without exchanges there are no trades and without trades there are no mutual benefit or meeting each other's needs. Then there

---

W. Zhao (✉)
Dalian Maritime University, Dalian, China

W. Liu (ed.), *China's Belt and Road Initiatives*,
https://doi.org/10.1007/978-981-13-0101-8_9

would be no place for the Maritime Silk Road strategy. Today, after 30 years of economic reforms and opening up, China has transformed from an inward economy to an outward one. By 2013, 90% of transport volume in China's foreign trade was accomplished via shipping. Cargo handling capacity of Chinese ports and the scale of China's shipbuilding industry are both the largest in the world. "In 2012, the size of China's maritime economy had surpassed five trillion RMB, accounting for 10% of the national economy. 90% of transport volume in China's foreign trade was accomplished via shipping. China was among the largest producers of marine oil and gas in the world. China also was the global leader in cargo handling capacity of ports, shipbuilding, marine-product farming and fishing[1]". Therefore, with its powerful shipping capacity, China is capable of building a comprehensive and interactive pattern of all-around opening-up, which links the entire continent of Eurasia and benefits the people along the coast. This pattern will also link seaports in China and those in countries along the Belt and Road, so as to realize the strategic goal of economic collaboration, cultural integration and shared prosperity. First of all, guided by the Shanghai Free Trade Zone, development of maritime economy is to be accelerated in ports such as Shanghai, Ningbo, Quanzhou, Xiamen, Guangzhou, Beihai, Fangchenggang and Qinzhou as well as the Yangtze River Economic Belt and the Pearl River-Xijiang Economic Belt; Secondly, with breakthroughs in key ports construction, connectivity through seaways will accelerate, so will the construction of large-scale infrastructure projects like strategic ports in the Indian Ocean such as Gwadar of Pakistan, Chittagong of Bangladesh and Hambantota of Sri Lanka. And construction of the China-ASEAN Free Trade Zone is to be upgraded.

The Maritime Silk Road strategy, driven by improved shipping capacity, has been widely recognized by relevant countries. Abdulla Yameen, President of the Republic of Maldives, upon fully realizing the significance of participating and supporting the Maritime Sild Road construction, has taken the lead in launching a series of maritime transport construction plans to fulfill the obligation as a coastal nation. The Maldivian plans include a cross-sea bridge linking Malé, the country's capital, to the island where its airport locates, as well as projects such as a large international port of transshipment in north Maldives and reconstruction of the Malé International Airport.[2] Lakshman Abeywardena, the Investment Promotion Minister of Sri Lanka, also thinks the Maritime Silk Road construction will provide a historic opportunity for the country's economic boom, facilitating Sri Lanka's rise as a global hub for aviation, shipping, energy, business, trade and tourism.[3]

---

[1]Director of the SOA: The historic opportunity for grand development of China's oceanic enterprise has arrived. www.people.com.cn November 7th, 2013.

[2]President of the Republic of Maldives: China has become Maldives' closes and firmest development partner. www.xinhua.net July 15th, 2014.

[3]An Exclusive Interview with Lakshman Abeywardena, the Investment Promotion Minister of Sri Lanka, www.gov.cn September 14th, 2014.

A container harbor

All in all, the maritime transport infrastructure like ports and seaways is the precondition for all economic and trade activities, so the foundation must be laid in advance.

Building of Maritime Silk Road Needs Guidance and Support from Policy and Law

Policy and law bear the functions of establishing rules and guiding social behaviors so their role as regulator and adjuster of social life is irreplaceable. As a long-lasting and large-scale project, building of the 21st century Maritime Silk Road involves dozens of countries and a hundred or so ports along the route. Therefore, only with guidance and support of policy and law can everything be carried out in a well-organized fashion. Countries along the route vary in political system as well as economic status.

Given the diversity in political, legal, economic and cultural backgrounds, though China has tremendous confidence in the Maritime Silk Road and has invested USD40 billion to establish the Silk Road Fund and to implement a series of financial measures, so as to demonstrate China's determination in building the Belt and Road. But, carrying on the Maritime Silk Road strategy requires more than just manpower and fiscal strength. Current relationship among countries and relevant policies on resolving conflicts must be taken into consideration as well. As an international economic and cultural strategic measure, it needs forward-looking overall planning and supportive policies to guide cooperation among countries and guarantee that future development won't go off-track from the original spirit of win-win. Naturally, the implementation of every policy needs to be guaranteed by proper negotiations and strict operating mechanisms. Lack of multilateral negotiations and consensus in terms of distribution of rights and obligations of governments and business entities along the route would only lead to failures in fulfilling stakeholders' multi-dimensional demands under a just and reasonable legal system. Building of the Maritime Silk Road would be stranded in frustration, too.

In other words, if there were any accidents or imbalances caused by inherited disputes over sovereignty or interest allocation among nations, new conflicts might be induced. Not only would this hamper economic exchanges, but it'd also interfere with normal international relations. Although "if properly managed, harbor trade is the most fruitful, with profit calculated by millions",[4] yet the groundwork for "harbor trade" must be laid by safe maritime transportation and friendly relationship between neighbors. Without those conditions, harbor trade is unlikely to be fruitful. China has had its share of hard lessons historically. During the Ming Dynasty, Chinese lives and properties were serious threat from Japanese pirates so Emperor Zhu Yuanzhang had to give the order that, as of December 1371, no Chinese individual was allowed to make contact with foreign countries.[5] In 1384, a further ban on marine fishing was announced.[6] These orders had considerably limited China's foreign trade thus severely impeded China's maritime economy. If the ban on maritime activities was a desperate choice under the circumstance, then, many years later, the lift of the ban and the establishing of a legal system with the state supervising and controlling maritime transportation could be viewed as a positive policy orientation. The Ming regime established a supervision system for maritime transportation with the Harbor Trade Bureau managing taxation and the government certification. This system fully demonstrated the crucial influence policies and laws have on maritime economy and trade. On the one hand, supervision on maritime activities was to guarantee safety for human lives and properties. On the other hand, it was another state policy for increasing revenue. As described by a scholar, the Ming Dynasty didn't officially ban all foreign trade or isolate China from the outside world, except for trade with Japan, which was never fully carried out, due to the issue of pirates, tribute trade with other countries remained active at all times. Besides tribute, private foreign trade, after the end of pirates during the

---

[4]Collection of Scattered Writings of the Collected Statutes of the Song Dynasty, Government Posts 44, Photocopy of the 1936 Edition, Capital Library of China.

[5]See Volume 139 of Veritable Records of the Emperor Taizu of the Ming Dynasty, Photocopy of the 1940 Edition, Library of Traditional Chinese Culture, Nanjing, Jiangsu; Source: An Feng's Studies of Forestry and Maritime Policies of the Ming Dynasty, Master's Thesis, Shandong University, 2007, Page 11.

[6]See Volume 159 of Veritable Records of the Emperor Taizu of the Ming Dynasty, Photocopy of the 1940 Edition, Library of Traditional Chinese Culture, Nanjing, Jiangsu; Source: An Feng's Studies of Forestry and Maritime Policies of the Ming Dynasty, Master's Thesis, Shandong University, 2007, Page 12.

Reign of Jiajing, was legal in the 100 years from the Reign of Longqing to the late Ming Dynasty.[7]

The Maritime Silk Road Strategy Require Countries Along the Route to Lift Their Ability of Performing International Conventions

E. Adamson Hoebel, the American legal anthropologist once wrote: "……the more civilized man becomes, the greater is man's need for law, and the more law he creates. Law is but a response to social needs.[8]" As an integrated project, building of the Maritime Silk Road must comply with certain laws and regulations. Rule of law is not only precondition and foundation of building the Maritime Silk Road, but also the guarantee for its sustainable development. Building the Maritime Silk Road isn't just exchanges among nongovernmental organizations but cooperation among countries. So, in that context, the term "rule of law" here doesn't refer to legal status of a certain nation or rules established according to multilateral treaties concerning several countries, but rather a set of co-created maritime legal order in accordance with international standards. Complying to international agreements is an inevitable choice when establishing order for maritime transportation.

The international community has launched a myriad of conventions on allocation of ship facilities and safety standards for ship crew. According to the principle of abiding by agreements, all contracting states shall strictly comply to relevant regulations covered by the convention. "To perform obligations covered by the convention or to implement rights stipulated in the convention, all contracting states shall take all appropriate measures in legislative, judicial, administrative and other sectors, as well as applying the negative inaction rule, which means not to take any action that's not in accordance with the convention.[9]" Otherwise, each country would have their own standards for ship safety and crew qualification, there would be no place for safety of seaways. As an advocator of the Maritime Silk Road, China should be the first to actively perform its international obligations. For international conventions approved by the National People's Congress, China should lead by example as performing obligations, in legislation, law enforcement and justice, as stipulated in the convention.

To ensure safety of seaways, inspection on maritime ships and crew qualification should be carried out according to 1966 *International Convention on Load Line*,

---

[7]See Zhao Yifeng's On Activities of Japanese Pirates in Hongwu's Reign, originally published on Ancient Civilizations, the 3rd Issue, 2013, Page 95.

[8]The Law of Primitive Man, by E. Adamson Hoebel, Translated by Yan Cunsheng, Law Press, the 2006 Edition, Page 6.

[9]The Application of International Criminal Treaties in China by Gao Xiudong, PHCPPSU, 2012 Edition, Page 40.

1976 *Convention Concerning Minimum Standards in Merchant Ships* and *1978 International Convention on Standards of Training, Certification and Watch-keeping for Seafarers*. In navigation, all ships should strictly abide by international conventions such as *1972 International Regulations for Preventing Collisions at Sea 1973 International Convention for the Prevention of Pollution from Ships, 1974 International Convention for Safety of Life at Sea, 1979 International Convention on Maritime Search and Rescue and 1989 Salvage Convention, so as to preserve safety and order at sea as well as guarantee a safe and clean oceanic environment.*

In recent year, occurrence of marine perils with considerable casualties increased globally. This has indicated how feeble safety management and supervision are at sea and also exposed severe insufficiency of salvage ability either of self or others. Article 98 of the *United Nations Convention on the Law of the Sea* especially stipulates the "salvage obligation" of captains, urging every country to instruct the captain of every ship with the country's national flag flown from it to "rescue any individual whose life is in danger at sea." The convention also requests every coastal nation to promote their "maritime search and rescue service" and "cooperation with neighboring countries in the region". To this end, China shall voluntarily bear the obligation for search and rescue at sea as well as unite with other countries along the coast in enhancing ability of search and rescue. "The ocean is ruthless, but people aren't", effectively limiting the casualty of marine perils is also an inevitable requirement for seaway safety.

The Maritime Silk Road Strategy Prompts Coastal Nations to Establish Safety Standards for Shipping

The Maritime Silk Road is to be connected by ports and put into use via shipping so universal standards and regulated management model would be an inescapable issue for countries along the route. Precisely, maritime transport requires coastal nations to follow systems that are in accordance with international industry standards, such as ship registration/inspection/visa, route administration, crew training and protection, marine rescue and salvage, international shipping insurance and taxation, marine conservation, etc. Only by complying to rules can human weakness be overcome thus reduce risk of human conduct, so our course can last and keep developing. For example, in marine conservation, it is necessary for Asia to learn from the EU and establish, as soon as possible, a prevention and punishment system against pollution from ships. Otherwise, the people along the route might lose their beautiful homes before they have a chance to enjoy the fruit of cooperation.

For the same ship-caused pollution of marine environment, results can be distinctively different in Asia and the EU. It is an enlightening phenomenon to be noticed. On June 13th 2001, a Malaysia-bound Indonesian tanker carrying 600 tons of toxic industrial phenol and 16 tons of chisel toppled near Malaysia's south coast, right across Singapore, causing death of all fish and seashells in eight adjacent farms. This incident was merely settled with civil compensation, whereas in 2002 in wake of the Prestige oil spill in Spain, the EU responded with measures that afterwards largely contributed to the improvement of European marine environment.

At the time, oil and fuel leaked from the Prestige spilled over a range of 300 km along the Costa Galicia, causing death of a great number of marine lives and unemployment of 120,000 fishermen. The Spanish government spent 42 million euros on the cleaning campaign, yet lives of hundreds of thousands of seabirds and otters were lost for good. This incident prompted the EU to launch a series of laws and regulations for the protection of marine environment, including: forbidding single-hull tankers to transport heavy oil; forbidding vessels flying flags of convenience to enter EU waters; Using a universal protocol when investigating maritime incidents; inflicting criminal punishment on parties at fault in pollution incidents. Regulations and orders listed above provide more than just a revelation but, more importantly, the establishment of rules for investigation on maritime affairs as well as interference of criminal and judicial authorities in pollution incidents by negligence.

2002 oil spill of Bahamian tanker prestige near Spanish coast

Seaway Safety Demands Joint-Guarantee from Law Enforcement and Judicial Authorities

John. F. Kennedy, former president of the US once gave a visionary speech that said: "To control the ocean is to have safety, to control the ocean is to have peace, to control the ocean is to have victory.[10]" As a matter of fact, as long as there are

[10]See Sea Power Theory by Alfred Thayer Mahan, translated by Fan Lihong, Shaanxi Normal University General Publishing House, 2010 Edition, Page 10.

human activities at sea, the issue of safety is inevitable. Dangers can come from nature itself, like tsunami, typhoon and sea tide, or they can be induced by human factors, like piracy, warfare and armed robbery. Fortunately, after the two World Wars, the 1945 *Charter of the United Nations*, in an effort to guide mankind out of wars and chaos and towards peace, stated clearly in the Preamble: "WE THE PEOPLES OF THE UNITED NATIONS DETERMINED to save succeeding generations from the scourge of war, which twice in our lifetime has brought untold sorrow to mankind, and to reaffirm faith in fundamental human rights, in the dignity and worth of the human person……to practice tolerance and live together in peace with one another as good neighbors, and to unite our strength to maintain international peace and security."

As a result, matters of traditional security, such as naval arms race, threats and wars at sea, have gradually sunk into oblivion whereas matters of non-traditional security such as marine environment safety, information security, smuggling and drug trafficking, crimes of terrorism, piracy, money-laundering and illegal immigrants, are deteriorating in the modern world. Merchant ships sailing along the maritime economic belt often encounter with said problems. For example, Pakistan is under constant threat of terrorism. In spite of its government and military carrying out massive strikes against terrorist facilities within the country's borders, Counter-terrorism and security situation in Pakistan are not getting any better. In 2011, a naval base in Karachi was attacked by shooters. During the standoff that lasted for 17 h, at least ten individuals were killed and two aircrafts were destroyed.

A total of 23 people, including security forces, were killed in the 2014 terrorist attack at Jinnah International Airport of Karachi. President Xi Jinping had been well aware of security situation along the Belt as early as he put forth the Maritime Silk Road strategy and had proposed an initiative of cooperation on disaster prevention and relief, network security, fighting transnational crimes and joint law enforcement operations. Rudolph von Jhering, the German jurist, once wrote: "The law is invalid if it's not mandatory" and "A law without mandate is like a fire not burning or a light not shinning." It's undoubtful that enhancing law enforcement and justice at sea is an effective way to deal with non-traditional security issues. Countries along the coastal economic belt should intensify preservation of maritime safety and order in the following aspects:

One, enhancing judicial cooperation on joint counterterrorism operations, especially punishing of piracy and armed robbery at sea, as well as taking special countermeasures. In recent years pirate activities have been transferring to Asia via Gulf of Aden. The number of international pirate activities on record peaked during January to May 2015, reporting a total of 80 cases including 75 committed crimes and five attempts, a 19% growth from the same period in 2014. In May 2015 there were totally 20 reported cases of piracy and armed robbery in Asia, increasing by

19% from April. Plus, most of these incidents took place in straits of Malacca and Singapore (SOMS).[11]

Though currently there is no specific charge against piracy and armed robbery at sea in the criminal law of China, relevant international convention does have a definite stipulation on punishing violent crimes at sea. According to article 9 of the Criminal Law of the People's Republic of China, "This Law shall be applicable to crimes which are stipulated in international treaties concluded or acceded to by the People's Republic of China and over which the People's Republic of China exercises criminal jurisdiction within the scope of obligations, prescribed in these treaties, it agrees to perform." Therefore, China can, in accordance with this convention, punish international crimes at sea, also known as the "adoption" rule.[12]

The 1982 *United Nations Convention on the Law of the Sea* explicitly stipulates components of piracy and relevant principle of criminal jurisdiction as stated in Article 105: "On the high seas, or in any other place outside the jurisdiction of any State, every State may seize a pirate ship or aircraft, or a ship or aircraft taken by piracy and under the control of pirates, and arrest the persons and seize the property on board. The courts of the State which carried out the seizure may decide upon the penalties to be imposed, and may also determine the action to be taken with regard to the ships, aircraft or property, subject to the rights of third parties acting in good faith." Article 14 of United Nations Security Council Resolution 1846 Calls upon all States, and in particular flag, port and coastal States, States of the nationality of victims and perpetrators of piracy and armed robbery, and other States with relevant jurisdiction under international law and national legislation, to cooperate in determining jurisdiction, and in the investigation and prosecution of persons responsible for acts of piracy and armed robbery off the coast of Somalia, consistent with applicable international law including international human rights law. Meanwhile, Article 15 notes that the 1988 Convention for the *Suppression of Unlawful Acts Against the Safety of Maritime Navigation* ("SUA Convention") provides for parties to create criminal offences, establish jurisdiction, and accept delivery of persons responsible for or suspected of seizing or exercising control over a ship by force or threat thereof or any other form of intimidation; urges States parties to the SUA Convention to fully implement their obligations under said convention…

China and other countries along the route, who are also contracting states of the *United Nations Convention on the Law of the Sea,* are obligated to perform their duties in punishing pirates. Furthermore, for armed robberies at sea other than piracy, the International Maritime Organization has a legal policy, too, as stipulated in Article 6 of the 1988 *Convention for the Suppression of Unlawful Acts Against the Safety of Maritime Navigation:* when the offence is committed (a) against or on board a ship flying the flag of the State at the time the offence is committed; or (b) in the territory of that State, including its territorial sea; or (c) by a national of

[11]The Worst in Five Years: A Steep Rise of Cases of Piracy and Armed Robbery At Sea in Asia, via Aiweibang, a Wechat App, June 12th 2015.

[12]See An Introduction to Law of Treaties by Li Haopei, Law Press, 2003 Edition, Page 516–517.

that State, each State Party shall take such measures as may be necessary to establish its jurisdiction over the offences set forth in Article 3, including: (a) seizes or exercises control over a ship by force or threat thereof or any other form of intimidation; or performs an act of violence against a person on board a ship if that act is likely to endanger the safe navigation of that ship; or (c) destroys a ship or causes damage to a ship or to its cargo which is likely to endanger the safe navigation of that ship; or (d) places or causes to be placed on a ship, by any means whatsoever, a device or substance which is likely to destroy that ship, or cause damage to that ship or its cargo which endangers or is likely to endanger the safe navigation of that ship; or (e) destroys or seriously damages maritime navigational facilities or seriously interferes with their operation, if any such act is likely to endanger the safe navigation of a ship; or (f) communicates information which he knows to be false, thereby endangering the safe navigation of a ship; or (g) injures or kills any person, in connection with the commission or the attempted commission of any of the offences set forth in subparagraphs (a) to (f). As of June 29th, 1991, the Standing Committee of the NPC officially approved the convention, authorizing and obligating China to implement its obligations under said convention.

Two, enhancing the ability of law enforcement and justice at sea, especially law enforcement collaboration with other countries, as well as perfecting the linking mechanism between administrative law enforcement and criminal justice. Maritime law enforcement and justice is a necessary tool of a sovereign state for controlling the ocean. In practice, due to differences in legal system and ideas, countries vary largely in strictness of maritime law enforcement and justice as well as their foreign policies. For example, on precisely 17:15 pm of December 17th, 2007, a super-sized Hong Kong tanker, the *Hebei Spirit,* which carried about 260,000 tons of crude oil, scratched and collided with a loose Korean floating crane, the *Samsung One,* while the latter was moored in the Port of Taesan-hang, Taean-gun, in the west coast of South Korea. The incident caused damage in three oil holds on the tanker's portside. Some 10,500 tons of oil spilled and polluted an area 7.4 km in length and 2 km in width.

As a result, the captain and first-mate were charged for neglect of duty and failing to performance their obligation of preventing the oil leak. The two were respectively sentenced one year six months plus a fine of 20 million Korean won, and eight months plus a fine of 12 million Korean won. Given the fact that the *Hebei Spirit* was scratched and collided by the *Samsung One,* the latter shall bear the primary liability. Nevertheless, the Korean authorities decided to put the captain and first-mate of the *Hebei Spirit* on a criminal court on the ground that they failed to stop the oil spill in time. Although the court's final decision was to exonerate both defendants, it at least showed that Korea was well-motivated in protecting the country's marine environment.

On the contrary, when similar cases occur along China's coast, the legal results are mostly civil compensation instead of criminal charges. For example, on May 11th 2007, a Hong Kong ship, the *Jinsheng,* which carried 239 containers, sailed from the port of Yantai. Around 3:00 PM of May 12th, it collided with a Korean cargo ship, the *Golden Rose,* causing an oil spill from the latter in adjacent sea area. After investigation, the *Jinsheng* and the *Golden Rose* were to respectively bear 55 and 45% of the liability. Ecological damage and loss of fishery resources caused by the oil spill added up to a total of 16.2 million yuan. The Shandong Province Ocean and Fisheries Bureau filed a lawsuit to the Qingdao Maritime Court and ended up with a compensation of 8.91 million yuan, paid by the liable parties, for the loss of marine ecological and fishery resources.

This 2007 incident was a tough trial duet to the fact that it involved a foreign party and compensation concerning marine ecosystem the result didn't come out until 2011. Within the industry, the successful trial of this case was considered an encouragement and guidance for demanding compensation of marine ecosystem damage as well as punishing pollution of marine environment. But in fact, this case was questionable compared to the *Hebei Spirit* case. For the same result of marine pollution, the Korean authorities put the defendant on a criminal court whereas their Chinese counterpart merely settled with civil penalty instead of criminal punishment, which is more effective in preventing crimes. One thing that's worth thinking is that for a long time China has overlooked the role of criminal punishment in controlling the ocean. There were even some scholars who proposed that sailors should not be held accountable because they bear greater occupational risk than most.

As the most severe of legal penalties, criminal punishment has a preventive effect matched by no other legal measures. We should realize that the deterrent force of criminal punishment lies not in its harshness but in its inevitability. At least, in matters of upholding state sovereignty and protecting marine environment, the educational and instructional function of the law should be reflected. We can't simply transfer problems of maritime safety to military deterrence or diplomatic relations.

Three, improving security controls over fleets in the Belt and the ability of accident prevention; Enhancing international criminal judicial cooperation and construction of the Asian-Pacific Maritime Justice Center; Cultivating members of maritime industries and law enforcement; Improving investigative skills in administrative law enforcement and criminal justice as well as the overall strength of law enforcement and justice system. To this end, China should enhance its maritime judicial ability and entrust Chinese maritime courts with the power of criminal adjudication.

Four, setting up a platform for collective security of the Maritime Silk Road; Installing and sharing maritime security surveillance facilities; Updating and sharing maps of nautical routes; Drawing on the experience of the US "Container Safety Protocol" (CS1), supervision on marine risks should come before containers leave the port; Tightening controls on security problems of ports along the route.

Five, in case of serious crimes concerning people's livelihood, such as piracy, armed robberies at sea and pollution of marine environment, relevant agreements should be signed to gradually regulate standards for incrimination, meanwhile minimizing disparities among nations to avoid malicious evasion of criminal liabilities.

Wei Zhao was graduated from the Law School of Renmin University of China. He is currently a professor and doctoral supervisor at the Law School of Dalian Maritime University, a Director of the Maritime Law Research Center, and a Director of the Dalian Maritime Law Research Base. His field of study is maritime criminal law.

Let the East Gale Unfurl the Lonely Sail—Historical Thoughts on the Modern Maritime Silk Road

Wang Jie

"The giant whale rests after many days a tranquil sea; Against a thousand sails the sun reflects as ships approach." As early as in Qin and Han Dynasties, the Maritime Silk Road, an ancient maritime transportation artery connecting East and West was established. It was one of the oldest maritime routes known to man. This seaway, which was the crystallization of wisdom and courage of oriental nations, has long borne the responsibility for cultural and commercial exchanges between East and West. Since the beginning of the 21st century, there have been frequent cultural and commercial exchanges among China and countries and regions along the ancient Maritime Silk Road. In light of this, the Chinese government seized the opportunity to put forth the strategic conception of building the 21st century Maritime Silk Road. The history mirrors both prosperity and depression. As for the issues of legacy and inheritance of the ancient route, I'm inclined to take a glimpse back in time to find some reference for the plan to rebuild the Maritime Silk Road today.

Going With the Flow of Time, the Millennia-Old Silk Road Revives: Development of China and the World Needs the Maritime Silk Road

Today, with the economic globalization, trade activities around the world have become firmly linked. No country can exist solely by itself. There would be no development for China without the world and vice versa. Economic development of China not only fuels economic growth of the world, but it also creates gigantic potential demand and a market of enormous scale for the world. In 2013, the total value of China's import/export was a staggering USD4.16 trillion, putting China on

W. Jie (✉)
Dalian Maritime University, Dalian, China

W. Liu (ed.), *China's Belt and Road Initiatives*,
https://doi.org/10.1007/978-981-13-0101-8_10

the number one position of the world. As the world's largest goods trader, China needs more comprehensive trades as well as a friendly relationship with the rest of the world. In the development of global trade that China participates, eastward economic exchanges are more mature.

After 30 years of China-US trade, ties between the two countries have increasingly deepened. China is now the US' second largest trade partner and the goods traded between the two sides have transformed from labor-intensive products to ones of technology and resources. Despite some turbulence in their relationship, Japan is still China's fifth largest trade partner. Export of capital/technology-intensive Chinese goods to Japan keeps growing. The composition of import and export commodities between China and Japan is gradually optimized. Suffice to say that both the quantity and the quality of Chinas eastward trade have reached a preferable state.

In comparison, China's westward trade is developing incessantly. There is space for growth in quantity and quality. As of 2013, total trade between China and South Asia, the Middle East and Africa is respectively USD96 billion, USD300 billion and USD210.2 billion. There is a huge gap between those numbers and that between China and its top ten trade partners (For example, USD559.04 billion with the EU). Meanwhile, in terms of commodity composition, China's westward trade seems rather flat. For example, with the Middle East, main commodities are oil and energy goods with only a rare share of high-value-added products. So, there is great potential in the quality of trade.

In October 2013, when addressing at the Indonesian Congress, President Xi Jinping put forth, for the first time, the idea of building the 21st century Maritime Silk Road. In the *2014 Report on the Work of the Government*, Premier Li Keqiang stated clearly that "We will intensify the planning and building of a Silk Road economic belt and a 21st century maritime Silk Road." Much like its ancient precursor, the 21st Century Maritime Silk Road will focus on the west. Starting from Chinese coastal ports, the route will stretch southward all the way through the Strait of Malacca and the Indian Ocean to reach Africa and Europe. To be more precise, building the 21st Century Maritime Silk Road is to build an economic belt of strategic cooperation on trade and shipping. It is a strategic move that aims to constitute a new perspective of comprehensive opening-up and propel the great rejuvenation of the Chinese nation.

Currently, the 21st Century Maritime Silk Road initiative is being carried out smoothly. Take port construction for instance, China has participated in many ports along the route, including Chittagong of Bangladesh, Gwadar of Pakistan, Colombo and Hambantota of Sri Lanka, Djibouti, Said of Egypt, Lagos of Nigeria and Piraeus of Greece. Construction of the following ports has started as well: Kyaukpyu of Myanmar, Jakarta of Indonesia and Bagamoyo of Tanzania. It is to be expected that a shipping network that links East, South and West Asia, as well as East and North Africa and even Europe, is gradually coming into being with ports along the route joining in one after another, facilitating wide cooperation among regions along the Maritime Silk Road and contributing to the realization of dreams for China and the world.

In a Time of Prosperity, Mankind Will Come Together Over Seas and Mountains: Abundant Connotation of the Maritime Silk Road

It is well-known that the term Maritime Silk Road has an ancient origin. As early as some one thousand years ago, ancestors of the Chinese and other nations had trudged across great mountains and oceans to open up the Maritime Silk Road, which then became a thoroughfare for trade and cultural exchanges between the two sides of the ancient world. It has promoted understanding and goodwill between China and foreign nations, as well as witnessed the maritime glory of ancient China. The Maritime Silk Road of the new age should inherit that legacy and old wisdom in order to build a road of peace, mutual-benefit, innovation and civilization, making its due contribution to the humanity.

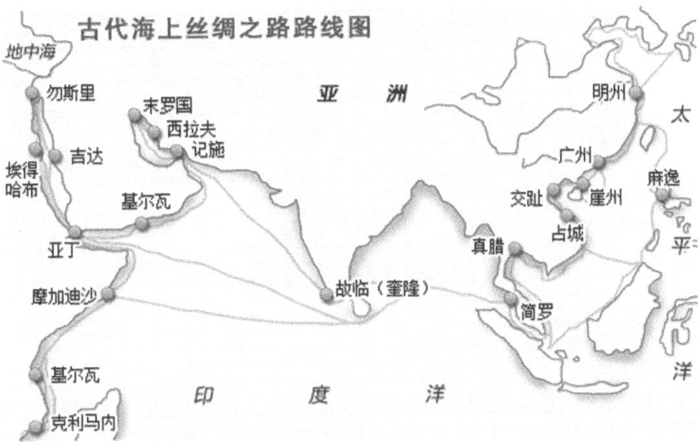

A route map of the ancient Maritime Silk Road

The Maritime Silk Road Is a Road of Peace

Ever since the ancient times, the Maritime Silk Road has been a road toward peace and friendship. Merchant ships from countries along the route such as China, Arabia, Persia and India were shuttling back and forth in peace and trading goods with each other. During the Ming Dynasty, Zheng He had made seven expedition trips to the western seas with a phenomenal group of over 200 ships and 27 thousand people. The group made several round trips to West Pacific and the Indian Ocean, visiting more than 30 countries and regions. This was a convincing example of the peace and friendship mentioned above.

In striking comparison, when the Europeans opened up new sea routes, what followed were blood, fire, violence, colonization and looting, which only brought destruction and catastrophe upon the locals. The Indian population in Haiti was once reduced to 500 from the original 60,000. In the 18th century, some seven million Africans were trafficked to America as slaves. According to the power logic of western colonists, Zheng He's fleet was more than capable of snatching territories overseas. Yet they did not take an inch of land from countries along the route. That's why it's called a group instead of a fleet. Another thing that's worth mentioning is that Zheng He's group once exterminated a gang of pirates in Southeast Asia and Malacca to help preserve safety and accessibility of the Maritime Silk Road, an act that had greatly benefited the people along the route.

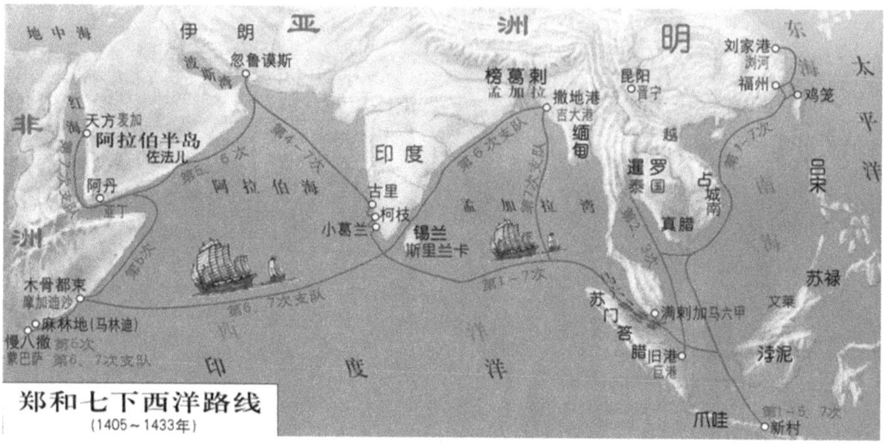

A route map of Zheng He's trips

Today, as we enter the 21st century, peace and friendship should be cherished even more when building the modern Maritime Silk Road. Over 40 years ago, governments of China and Tanzania co-founded the Chinese-Tanzanian Joint Shipping Company, which had made tremendous contribution to the construction and operation of the TAZARA Railway. Up to this day, the company remained the one and only international shipping company in East Africa. The people of Tanzania and Zambia, and the entire Africa as a whole, all considered the TAZARA a railway of freedom and an excellent example of South-South cooperation. Naturally, when building the 21st century Maritime Silk Road, we should carry on that grand tradition and fully realize that the Maritime Silk Road should be a road of peace and friendship instead of one of dominance-seizing. We should comply to the principles of equality, mutual-benefit, peaceful use and shared development when building the 21st century Maritime Silk Road, the road of peace.

The Maritime Silk Road Is a Road of Mutual-Benefit

Mao Zedong once said that China has a long coastline that provides us with convenience of communicating with all nations overseas. About one thousand years ago, ancient China had established equal and friendly trade relations with many foreign countries, benefiting the people along the Silk Road. During the Song and Yuan Dynasties, when the Maritime Silk Road was in its prime, Chinese merchants sailed out to the open sea, delivering goods like silk, porcelain and tea to Southeast Asia and India, then brought back spices, drugs, etc. Not only did the Chinese civilization spread to countries overseas, but the people of China also benefited from the trade. During the Southern Song Dynasty, tariff from foreign trade even accounted for one seventh of the government's total revenue. The seaways were wide open and the goods were transported unhindered. It was a win-win situation where countries got what they needed and benefited from each other. On a different note, extensive discoveries of new routes in the 15th century marked the beginning of maritime looting by western colonists. From 1480 to 1530, the Portuguese snatched ten percent of the world's gold from the Gulf of Guinea alone. From the 16th to the 19th century, Spain obtained 2.5 million kilos of gold and 100 million kilos of silver from the Americas. European colonists gathered themselves a massive amount of wealth at the cost of countless Indian and African lives. After the new routes were opened, many nations of Asia, Africa and the Americas slowly became colonized or semi-colonized. A distorted economic relationship, which was built around maritime powers of western Europe with unfair and unjust shipping systems and trade patterns, was established between the suzerain states and their colonies.

In recent years, China has been zealously promoting building of the Maritime Silk Road by constructing a great many of ports in South Asia for the purpose of connectivity. China directly invested USD1.4 billion in the construction of Colombo Harbor. Chinese construction projects in South Asia not only helped complete local infrastructure but also boosted development of portside industries. Plus, the harbor, once constructed, will attract new sources of goods and enhance commercial exchanges between the two sides as well as guarantee accessibility of China's energy corridors thus increase profit for Chinese enterprises. As a result, it will bring shared prosperity to all countries along the route. Therefore, the 21st Century Maritime Silk Road is going to be a road of mutual benefit. We should inherit the grand tradition by upholding the principle of mutual benefit and accomplish this maritime passage that profits both China and the world.

The Maritime Silk Road Is a Road of Innovation

The ancient Maritime Silk Road gave birth to many innovations and inventions that had a lasting influence on human civilization. Compass, one of the four great inventions of ancient China, when used by seafarers, had directly facilitated crucial development of navigation and maritime civilization. It is more so with modern-day Maritime Silk Road. The 21st Century Maritime Silk Road, relying on the Asia-Europe sea route, one of the three major sea routes of the world, with a total length of 11,000 nautical miles, possesses the typical features of "long routes-large ships-big ports". With building of the 21st Century Maritime Silk Road furthers, innovations are likely to boom in vessels-manufacturing, forms of shipping companies and development of ports. Shipping and trade on the route will thrive even more.A route map of Zheng He's Trips

A model of watertight bulkheads

Speaking of vessels of ancient China, many would think of Zheng He's ships. It was with the prosperity of the ancient Maritime Silk Road that Chinese ship-building techniques were greatly upgraded. Watertight bulkheads used on Chinese merchant ships were only adopted by other countries till 600 years later. The world's earliest balanced rudder was discovered on the Song Dynasty wooden ships in Jinghai, Tianjin. Entering the 21st century, the modern-day Maritime Silk Road, basing on the Asia-Europe sea route, which is the longest sea route in the world, starts from East Asia and stretches across the Strait of Malacca, the Indian Ocean and the Suez Canal to reach European ports. A stunning distance of 11,000 nautical miles with ships of large and super sizes sailing freely. Take Maersk for instance, as a giant of global shipping industry, the group now has two 18,000-teu jumbo ships operating on the Asia-Europe route. With further connectivity of the Mari Time Silk Road in the future, demand for ultra-large vessels will expand thus gives impetus to innovation of vessel-related technologies.

Ship companies, as the main force in connectivity of the Maritime Silk Road, are able to accomplish much in innovation. It is well-known that the Maritime Silk Road covers a variety of countries and regions that have abundant resources and ample commodities, providing an excellent experiment field and vast marketplace for shipping companies to innovate. All the major liner operators such as Maersk Shipping, CMA, MSC, CKYH Alliance and CSCL have a significant allocation on the Asia-Europe route. The companies keep innovating with business models by operating in various forms like single or joint venture and bulks for rent or sale.

Meanwhile, with densely scattered routes, the Maritime Silk Road is leading the innovation of shipping networks. The old forms of base ports and non-base ports have been broken and new ones like pivotal ports, main ports, branch ports and even feeder ports are occurring one after another. The double-layer shipping network will be replaced by triple or even quadruple-layer networks.

As bases of the Maritime Silk Roads, ports must be able to innovate too. Historically, ports were built in accordance with prosperity of shipping activities. As one of the starting points on the ancient Maritime Silk Roads, Guangzhou has been one of the largest ports in the world since the Tang Dynasty. Around 777 A.D, some 4000 ships arrived at Guangzhou annually. The Foreign Trading Bureau, China's first ever maritime trade administration, was established in Guangzhou. As time rolled on, in the 21st century, the modern-day Maritime Silk Road is largely based on the Asia-Europe sea route, which possess the typical features of "long rout-large ships-big ports". Therefore, stricter requirements have been introduced in terms of the scale of ports, handling efficiency and management. Against such background, major shipping companies like Maersk, Evergreen and COSCO have transformed themselves into the biggest port operators in the world, adding fuel to innovation of ports.

With the building of the 21st Century Maritime Silk Road advances, it is to be expected that countries and regions along the route will put up an even stronger growth in innovation. And the 21st Century Maritime Silk Road will become literally a road of innovation.

The Maritime Silk Road Is a Road of Civilization

As is known to all, transmission of religions in the ancient world followed the same path as commerce and trade. Buddhism or other western religions were all closely linked with the ancient Maritime Silk Road. The thousand-year-old Mosque in Quanzhou, Fujian, the 700-year-old Christian monument and Manichean temple… it was the Maritime Silk Road that made Quanzhou a hub of eastern and western religions in the ancient world. Obscure merchants of different nationality and faith sailed along the route. It was through the Maritime Silk Road that the torches of eastern and western civilization were passed on, bringing different values and civilizations together to form a trade exchange system of equality and friendship between east and west. Ibn Battuta, the great Moroccan traveler came to China by sea and stunned the Islamic world with his *The Travels*. The Italian explorer Marco Polo once led the convoy fleet of Yuan Dynasty escorting a Mongolian princess to Persia for a royal marriage. The entire Christian world was amazed by *The Travels of Marco Polo*. It was with the compass, a Chinese invention, and imagination of the orient inspired by *The Travels of Marco Polo,* that Christopher Columbus finally discovered the new continent, and ergo, the mistakenly naming of the East Indies and the Indians. However, after Columbus opened up the new route, the "road of civilization" concept of ancient east didn't get carried on. Instead, colonies were forced to depend largely on the their so called suzerain states who rose rapidly through exploiting and plundering weaker nations then used them for supplies of raw materials via sea routes and maritime trade. Eventually, an unreasonable system of international economy and trade was established. Even to this day, numerous African countries still suffer from singularity of economic structure. Some merely rely on the export of one or two raw materials so they have to be overwhelmingly attached to the international market. For example, in 1989, Ghana's coco production grew by 58% from last year yet their export earnings were down by USD65.2 million. The reason was that the price of coco in international market had decreased to USD1000 per ton from USD1800 in 1988. This was just the tip of the iceberg that was the consequence of the kind of "civilization" brought by new routes.

The modern-day Maritime Silk Road, as the "road of civilization" with a long history, is an important instrumentality for building a new prospect of global economic openness under new circumstances. In the future, with further progress of the Maritime Silk Road, a series of new public discourse are bound to emerge. Issues like rights distribution in the global shipping industry or monopoly shipping organizations have yet to be solved. On the one hand, for too long have western nations dominated the rule-making of shipping. Current international shipping standards are almost entirely set by the west. On the other hand, international cartels, the notoriously monopolistic organization worldwide, are still operating in the shipping industry disguised as liner guilds. Western economists even called them "pirates of the modern times". With the Maritime Silk Road serving as a platform, countries along the route can demand for equality in matters concerning interest, thus help to partly change the unfairness. Suffice to say that building the Maritime Silk Road involves both material and spiritual dimensions. Trade is not only a barter but the interaction and exchange of cultures. Therefore, the "road of civilization" shall be the path taken when building the Maritime Silk Road and this is the essential meaning of the modern-day Maritime Silk Road.

With a Well-Laid Foundation, the Time for an Epic Voyage Will Come—It Is the Perfect Time for Building the Modern-Day Maritime Silk Road

The Chinese began the exploration of oceans long ago. From Xu Fu's three voyages in the Qin Dynasty to Zheng He's seven expeditions in the Ming Dynasty, ancient China's maritime adventures had always been a magnificent chapter in the history of navigation. Since the 1990s, China's development has reached a new height.

Consistent growth of foreign trade and the strength of shipping industry as well as progress in ship-building techniques combined to provide a solid foundation for the development of modern-day Maritime Silk Road.

Trade is the primary driving-force for the development of navigation. China has a long history of foreign trade. During the Yuan Dynasty, with the establishment of ancient Maritime Silk Road, China developed a maritime trade relationship with nearly a hundred countries and regions, including those in the West Pacific and the Indian Ocean. During the Ming Dynasty, Zheng He's legendary journeys had pushed China's official trade to its peak. Since the economic reforms, China's foreign trade has been growing rapidly. The total import-export volume surpassed four trillion dollars in 2013, marking China's overtaking the US as the largest trading nation in the world. Countries and regions along the 21st Century Maritime Silk Road are China's major trade partners. Frequent trade exchanges with countries and regions along the route have laid a solid commercial foundation for building the modern-day Maritime Silk Road.

Trade relies on shipping. China has a long history of navigation. The ambassador ships in the Han Dynasty, the foreign trade route of Guangzhou in the Tang Dynasty and Zheng He's expeditions in the Ming Dynasty, those were all epic achievements in the navigation history of man. Prosperous civilian maritime activities gave rise to cosmopolitan ports in coastal China such as Guangzhou, Quanzhou, Mingzhou and Yangzhou, names of which can be found in foreign documents like *Ancient Accounts of India and China, The Travels by* Ibn Battuta and *The Travels of Marco Polo.*

Now, entering the 21st century, China's shipping industry once again embarks upon a promising future. The handling capacity of Chinese ports have been leading the world for eleven consecutive years. Eight of the ten largest ports of the world are based in China. China also has seven of the ten largest container ports in the world. Collective transport capacity of Chinese shipping industry has reached 140 million DWTs, accounting for 8% of the world's total capacity, the 4th largest globally. COSCO, China's biggest shipping enterprise, has the second largest operating fleet in the world. The consistent growth in shipping capability has laid a solid foundation for China when building the 21st Century Maritime Silk Road (Table 1).

The foundation of shipping is ship-building. Having a long history of building ships, China used to lead the world in ship-building technologies. According to *Ancient Accounts of India and China*, Chinese ships in the Tang Dynasty were so large that they were the only ones that could withstand the violent waves of the Persian Gulf. Zheng He's treasure ship was 148 meters long and 60 meters wide. With nine masts and twelve sails, it was the biggest sailing boat in the world. The three major types of ancient Chinese junks (Fuchuan, Guangchuan, Shachuan) were among the most brilliant inventions in the history of navigation. Entering the 21st century, China's ship-building industry reclaimed its place at the top. In 2010, China overtook South Korean as the world's largest ship builder, with the ability to independently develop mainstream vessels. Also, China has made breakthroughs with high-tech/high added-value ships and marine engineering equipment,

Table 1 Ten world largest ports by handing capacity in the year of 2013

Rank	Port	2013 (hundred million tons)	2012 (hundred million tons)	Year-on-year growth rate (%)
1	Ningbo-Zhoushan	8.0978	7.44	8.80
2	Shanghai	7.7600	7.36	5.50
3	Singapore	5.5800	5.38	3.70
4	Tianjin	5.0100	4.76	5.00
5	Guangzhou	4.5512	4.34	4.87
6	Suzhou	4.5430	4.28	9.30
7	Qingdao	4.5000	4.02	10.60
8	Tangshan	4.4620	3.64	22.40
9	Rotterdam	4.4110	4.42	−0.20
10	Dalian	3.3340	3.74	10.10

providing, along with other progresses listed above, the technological guarantee for building the 21st Century Maritime Silk Road.

The foundation is laid and the path is open. China, with a splendid legacy of its thousand-year-old navigation culture and the opportunity offered by the 21st century, will compose a new brilliant chapter of modern-day Maritime Silk Road.

Pursuing Greatness, Against a Thousand Sails the Sun Reflects as Ships Approach—The Modern-Day Maritime Silk Road and Its Abundant Cultural Connotations

The influence of culture on a society is subtle as it is prolonged. It is the same for navigation. Chinese navigation culture dates back to the Qin and Han Dynasties, with its uniqueness gradually taking shape throughout history. It is a culture of harmony, exploration, practicality and substantiality. Its influence on the modern-day Maritime Silk Road is not to be overlooked.

The harmonious spirit runs through the history of navigation in China and has been the ever-lasting theme. The harmonious characteristic of Chinese navigation culture originated in a unique faith. Comparing to Poseidon, the capricious irritable sea god in western mythology, Mazu, the mother goddess of the ocean in Chinese folk culture, is a symbol of hope and peace, reflecting the people's dream of a safe and tranquil sea environment as well as Chinese people's wish to maintain a friendly relationship with the ocean. As legend goes, Mazu used to be a woman of the Lin family in Meizhou Island, Putian, Fujian. She was a great swimmer who often helped fishermen in danger. Over time, she was apotheosized as a goddess at sea and later her divine status became official. Imperial governments since the Song Dynasty had, for several times, conferred on Mazu titles such as *Lady, Heavenly Princess, Heavenly Queen and Sacred Mother in Heaven.* A prominent symbol of

Chinese navigation culture, Mazu's image evolved with the development of ancient Maritime Silk Road. Not only did Mazu have a profound influence on other navigation cultures, but also provides a proud cultural foundation for building the modern-day Maritime Silk Road.

Rooted deeply in Chinese navigation culture, the exploring spirit is a driving force for enterprising on the Maritime Silk Road. Chinese navigators were courageous in their exploration of the ocean. As early as 2100 years ago, in the Han Dynasty, China opened up in the Indian Ocean the first long-distance sea route, which stretched for over ten thousand kilometers. This route, which took ships then more than a year to cover, was established some one thousand years earlier than the one discovered by western nations. Another important fact was that violent environment at sea never stopped the Chinese pioneer navigators, who risked their lives setting out to the big blue, facing unforgiving open waters of the Pacific and the Indian Ocean. Xu Fu, who set out on a voyage of no return in the Qin Dynasty; Jian Zhen, who lost his eyesight sailing to Japan during the Tang Dynasty; Zheng He, who died in a foreign land after having accomplished seven epic expeditions in the Ming Dynasty… these trailblazers of Chinese navigation wrote their heroic stories with exceptional insight and boldness. To quote Lu Xun: "We have seen countless devotional or fearless souls in our history and they shall be the backbone of the Chinese nation." Today, we should take up the torch of exploration and, with paramount bravery and strength, make every effort to build the modern-day Maritime Silk Road.

Practicality is another distinct characteristic of Chinese navigation culture. Practice at sea was the reason why ancient Chinese navigation technology outshined other countries in East Asia. Through years of maritime activities, Chinese navigators managed to record their sophisticated experience. For example, *Zheng He's Nautical Charts* was the most practical reference book for sailing in the world during the first half of the 15th century. It combined multiple functions like sea map and route guide and everything one needs to know about sailing. Building the modern-day Maritime Silk Road is a mega-scale systematic project far more complicated than anything in Zheng He's time. That's why we must stay earnest and down-to-earth and build the Maritime Silk Road of the new era with a pragmatic attitude.

Substantiality is also significant in Chinese navigation culture, which had a profound tie with trade in goods. Transmission of navigation culture and development of maritime trade were mutually-supplementary. During the Tang and Song Dynasties, the ancient Maritime Silk Road was in its prime. China exported tremendous volumes of silk, porcelain and ironware and in return imported pearls, logwood and lilac from overseas. It was a time of thriving navigation culture, which brought about the official ceremony praying for wind to aid sailing ships as well as Mazu worship.

Along with commodity trade, Chinese navigation culture crossed great oceans to reach foreign lands along the ancient Maritime Silk Road, laying in advance a solid foundation for international cooperation on the Maritime Silk Road of the new era.

The subtle influence of Chinese navigation culture provides an enriched cultural connotation for building the 21st Century Maritime Silk Road while its inclusiveness determines that the new route, which will be built upon mutual benefit and peaceful development, is a decisive answer to the call of the times and is in accordance with the people's demand for development.

Carrying on Past Traditions With a Fresh Start, There Will Be an Even Brighter Future: Path Choosing for the Modern-Day Maritime Silk Road

Sharp tools make good work. With the rise of trade, shipping and ship-building industry of China, basic conditions for building a shipping network are now met, providing the material basis and platform for development of the 21st Century Maritime Silk Road.

From a historical point of view, establishment of the ancient Maritime Silk Road was carried out gradually and geography-wise due to technological limitation. Back in the Han Dynasty, when the Maritime Silk Road first came into being, the farthest point it reached was the southeastern coast of India. And during the Sui and Tang Dynasties, Chinese merchant ships had sailed beyond the India Peninsula and into the Persian Gulf. In the Yuan Dynasty, Chinese merchant ships sailed even farther to enter the vast area off the west coast of the Indian Ocean. It was till mid and late Ming Dynasty, when the Europeans were sailing east, that the ancient Silk Road finally came to an end.

Today, thousands of ships flying the Chinese national flag are sailing on the routes from East Asia to the South China Sea, the Indian Ocean as well as Africa and Europe as they continue to write the interrupted glorious history of the Maritime Silk Road. With trade exchanges becoming more frequent and vigorous between China and Southeast Asia, Africa and Europe, a tight and invisible net of maritime trade is gradually linking countries together. So, we should inspect the Maritime Silk Road of the new era from a network perspective. Meanwhile, rapid development of navigational technologies ensures that the 21st Century Maritime Silk Road won't be limited as its ancient precursor. Therefore, when building the 21st Century Maritime Silk Road, the path we take shall be one that is network-based, geographically-planned and thoroughly-developed with breakthroughs in encompassing countries along the route. Centered on the establishment of the shipping network and based on trade exchanges between China and ASEAN countries, enhancing the development of shipping and trade ties with countries in South Asia, East Africa and the Middle East and eventually reaches Europe, the home of Marco Polo, for the road to be thoroughly developed. The shipping network of the Maritime Silk Road, in particular, must be woven with "needles and thread". "Needles" refer to ports, which are the foundation of the network. And "threads" refer to routes connecting those ports, which are the core of the network.

When plenty of ports and routes interweave, they form a comprehensive multi-dimensional shipping network, only with which can the modern-day Maritime Silk Road be built efficiently and effectively.

Gazing into the future, with peaceful rise of the Chinese nation as well as united efforts of the international community, the great strategic conception of the 21st Century Maritime Silk Road will definitely come to be realized step by step, paving a splendid way to the Indian Ocean and the world as well as making a significant contribution to the great rejuvenation of the Chinese nation while painting a dreamy picture for the people of China and the world.

Wang Jie Professor and doctoral supervisor at the Transportation Management College of Dalian Maritime University Field of Expertise: Developmental Strategy of Transportation, Planning and Management of Transportation, International Shipping Policies.

Diversity of Chinese Culture and the Belt and Road Initiative

Shiyuan Hao

At the 2014 Central Work Conference on Ethnic Affairs, President Xi Jinping pointed out that ethnic regions are rich in resources with sources of many river systems originated there. These regions serve as ecological barriers and have unique cultures. These are also regions on the border and are suffering from poverty. Only by acknowledging the basic facts can we truly understand China's national conditions and strategic significance of ethnic affairs so that we can be confident of practicing systems, laws and policies designed to solve ethnic problems.

In the past, we used to know China as a country of sheer size and huge population with abundant resources but stranded in poverty. But cultural uniqueness is an important symbol of ethnic and border regions that are part of the colorful collage of Chinese culture and epitomize China's characteristic diversity. Whether by observing the cultural sector from the culturological point of view or according to China's cultural composition including the linguistic system, the cultural diversity of the regions on our land borders is noticeable. Furthermore, not only do these regions have diverse cultures, but they also have diverse landscapes and biological systems. Certainly, most of these places are under-developed and impoverished, too. Given these basic conditions, especially with China being a unified multinational state, we need to have a good grasp and understanding of the "properties" of this big family.

Today, when pursuing the Belt and Road initiative, a significant change has taken place in western China, especially ethnic regions on the border. That is, they have transformed from places left behind in economic reforms and opening-up to the frontier, which opens both domestically and internationally. This critical change, which possesses historical significance, will have a lasting impact that is profound and meaningful. The diverse cultural resources of western China, especially border regions inhabited by ethnic minority groups, will become an advantage in China's all-directional opening-up. Making good use of this "soft power",

S. Hao (✉)
Mongolian Academy of Sciences, Ulaanbaatar, Mongolia

W. Liu (ed.), *China's Belt and Road Initiatives*,
https://doi.org/10.1007/978-981-13-0101-8_11

respecting differences and tolerating diversity, those are the keys to well-managed domestic affairs, which include peace and unity among all ethnic groups as well as harmonious development, and also, the fundamental principles of diplomatic practices such as amicable relationship with neighbors, mutual-benefit and win-win cooperation.

Basically, China's Belt and Road Initiative proposal is to comply with the trend of multi-polarization, economic globalization, diverse cultural development and social informatization, and most importantly, to uphold the principles of harmoniousness and inclusiveness. It has come to our notice that *Vision and Actions* stressed the following ideas in particular: civility, inclusiveness, respecting other nations' choices pertaining to development, enhancing dialogues between different civilizations, seeking common ground while reserving differences, inclusiveness, peaceful co-existence, shared prosperity.

This is in fact a cultural perspective, which means, when understanding culture in the broad sense, we must understand its diversity, and more importantly, how different cultures co-exist and share prosperity. The path and mode of a country's development also fall under the category of political culture and should be respected, accepted, tolerated and understood. The unique socialist path that China has taken isn't necessarily incompatible with western capitalism insofar as the two can't co-exist or share prosperity. China does not and will not inflict our choices upon others, nor will we accept condescending gestures from the US or other western nations, let alone following others in a groveling manner.

From a political-culture point of view, both sides should be capable of seeking common ground while reserving differences and agree to disagree and live in peace with each other. In that sense, the demonstration effect of ethnic regions is an important cultural advantage because opening-up itself is not just limited to economy but it's related to culture as well. Demonstration effect and cultural advantage of ethnic regions on the border are extremely important. Next, I'll address issues in three aspects pertaining to national conditions and two-way opening-up.

Origin and History of Diversity of Chinese Culture

According to documents of the pre-Qin Period, "tribes of the five directions" shared governance of China. The five tribes in a world with "round heaven and square earth" were, respectively, Yi of the east, Man of the south, Rong of the west, Di of the north and Huaxia of the middle. In this political structure centered on the middle region, the Middle-Kingdom's traditional policy on the four "barbarian tribes" could be summarized as "tolerate their beliefs and respect their customs, govern them without changing their traditions." To rule according to their habitude and pursue harmonious co-existence. As President Xi Jinping put it: "This idea of preserving unification while respecting differences is of vital importance to the forming and development of the Chinese nation."

The history of China is a history created by interaction and amalgamation of "tribes of the five directions". China will stay a multinational state no matter it's ruled by which people. In fact, the more powerful a dynasty was, the more peoples it included. Whichever people that succeeded in ruling China saw themselves as the legitimate ruler of a unified multinational state. And that is the main theme under which chronicles and dynastic history of ancient China were written. Our ancestors didn't stop writing the history or alter it because the ethnic identity of the previous dynasty. Grasping the law of development for Chinese history as well as forming and development of a unified multinational state, acknowledging the fact that China was built by "tribes of the five directions", those are views of Marxism and attitude of historical materialism.

In terms of cultural diversity, the first thing we must clarify is that mankind is unified but culture is diverse. Homo sapiens migrated out of Africa some 200,000 years ago and scattered themselves all over the world, replacing all previous humans of different evolutional origins. So, mankind is unified in nature and there is no disparity within this unification. Race is but an anthropobiological feature that has no cultural connotation. Therefore, there is no such thing as a racial culture or language. People are the carriers of culture and cultural diversity is decided by ethnic diversity. Diversity means differences and differences are the most common phenomenon in any human society. Of course, differences also mean conflicts and conflicts are all about how we see differences among individuals, peoples or even countries, all can be summarized as our cultural perspective or, in a larger sense, civilizational perspective.

Biological diversity is the very foundation of ecological balance. This is a basic scientific fact known to all. Then, why can't cultural diversity become the foundation of a harmonious society? Why must different cultures confront each other to an extent that it becomes civilizational conflicts? Those are some of the problems we must consider and solve when studying culture and diversity.

Evidently, minority cultures are the most prominent features of Chinese culture. China has 38 UNESCO-approved intangible cultural heritages, among which 14 are from minority cultures. The State Department announced three batches of national intangible cultural heritages respectively in 2008 and 2011. Among the 1219 national intangible cultural heritages, 433 are from minority cultures, accounting for 34.9% of the total. Among the 1986 inheritors of national intangible cultural heritages, 524 are ethnic minorities. In 2014, the State Department announced the 4th batch of national intangible cultural heritages and among the 153 subjects, one third are from minority cultures. Those examples give us a glimpse of the sheer proportion of minority culture in China's cultural diversity as well as relevant preservation and inheritance.

Meanwhile, what we must realize is that, while the data reflects the fact that minority cultures contribute so much to the abundance and colorfulness of Chinese culture, preservation, inheritance and development of minority cultures are imperative. Just as President Xi Jinping pointed out, minority cultures aren't strong enough to withstand the impact of market economy and some intangible cultural heritages have been lost so we should know better to cherish them before the day we lose everything. Minority cultures belong to all Chinese people instead of any one ethnic group or area. To show our respect for cultural diversity, the first thing we must do is to protect the cultural diversity of China. Chines culture is an integration of every ethnic culture. To say that Han culture is Chinese culture or to ignore minority cultures, as well as exclude any ethnic culture from Chinese culture or refuse to identify with Chinese culture are all incorrect thinking that must be overcome.

The long-term goal and fundamental principle of enhancing national solidarity is to reinforce cultural identity and build a shared spiritual home for all ethnic groups as well as promote the awareness of China being a community of shared nationality. Cultural identity is the most profound identity as well as the heart and soul of national solidarity and harmony. So called national solidarity and harmony simply refer to the root of Chinese culture. We should all identify the Chinese nation and its culture. Only by identifying Chinese culture can we identify the state, socialism and the Chinese nation. So, the role of culture is one of profundity and spirituality. Only by establishing cultural identity can we consolidate our identification of this great country, the Chinese nation and the unique socialist path China has taken.

The Identification of Chinese culture is shared by all ethnic groups and includes each group's self-identification. It leads to mutual-identification among different peoples. That is to say, each member of this diverse and unified family is a member of the Chinese nation whilst keeping their own names. If one only knows one's first name but not the family name, or only knows one's own people but not the big family they belong to, there will be misidentification, in which case one would only know oneself but not others or even sees oneself as above others so one would be conservative of self and rejective of others, sinking into the trap of narrow identification.

In that sense, multi-nationality is an advantage. And it's reflected not only in culture but also in the "properties" of the family, including advantages such as abundant resources, sources of river systems and ecological barriers. So, we must cherish this unified multinational state or consider ourselves lucky for being a unified multinational state with a long history. Therefore, when dealing with issues of ethnicity or cultural differences, we'd get nowhere if we treat ethnic diversity as a burden or ethnic issues as troubles, or even, treat ethnic minorities as foreigners and try to solve ethnic issues once and for all by abolishing ethnic identities or simply ignoring their existence. There were no successful precedents to follow anywhere in the world, instead, there are plenty of examples of transformation from denial to recognition. For instance, in 2014, the British government officially recognized Cornish people as a national minority group same as the Scottish, the Welsh and the Northern Irish.

We've been stressing the fact that China is a unified multinational state and the term "unified" refers the unification of the Chinese nation. Only with a unified Chinese nation can there be a unified state with territorial integrity. And "multi-national" refers to the historically existed diversity and differences among ethnic groups in language, culture and customs. To recognize and respect the differences is the attitude of historical materialism and a precondition for a harmonious relationship among all ethnic groups.

China's basic policy on ethnic issues is to respect differences and narrow disparities. It is a principle established from both spiritual and material perspectives. It is also an important reflection of the Chinese ethnic view on ethnic matters. Narrowing disparities means narrow disparities in economic an social development. Respecting differences means to respect cultural differences in the broad sense. Both are the essence of China's ethnic policy. This respect for differences and the sense of inclusiveness are largely based on the ideas of seeing harmony in differences and seeking unification in diversity. We are to enhance the national solidarity and establish a spiritual bond of equality, solidarity, mutual-help and harmonious relationship among all members in our diverse and unified big family, in other words, our shared spiritual home.

Generally speaking, Chinese culture is the co-existence and symphony of all ethnic cultures. It is like in an orchestra where sounds of various instruments and tones, coordinated by the conductor, unite and symphonize to create one beautiful harmony. "Apply the five-tone scale to the eight natural sounds, in coordinating and harmonizing them one shall find the way to rule a country." This is what our ancestors learnt from harmony in music. In that sense, Chinese culture is more like a modern symphony that consists of each and every ethnic group and not one less. Only by performing the cadenza of the Chinese nation in a symphony of diversity can we achieve this diverse unification based on respect for differences.

This modern "symphony" of Chinese culture includes tunes and tones of all ethnic cultures. With the Communist Party as the "conductor" and the core values of Chinese socialism as main theme, the cadenza of the great rejuvenation of the Chinese nation is being performed. This is the identification of Chinese culture, which is based on respect for differences as well as diversity and unification. And this is the spiritual convergence of the Chinese people.

Being Determined and Confident in the International Community

The 18th CPC National Congress put forth the doctrine known as the "Three Confidences", namely confidence in our chosen path, confidence in our guiding theories and confidence in our political system, which are the self-awareness we must reinforce when pursuing socialism with Chinese characteristics and building a moderately prosperous society in all respects.

In recent years, discussion on China's ethnic policy is an important subject for international intellectual community and the general public. And their basic conclusion is that there are problems with China's ethnic policy, so they, in so called "introspection", came up with a series of judgmental labels such as "the Soviet pattern", "a miscellany", "in need of adjustment" or "total failure". Meanwhile, a series of prescriptions based on their pseudo-proposition of "depoliticization of ethnic issues" were announced: the "cultural melting pot" theory taken from the US, India and Brazil; theories of "culturalization", "ethnic groups" and "social policies". "The clan system" of the Republic of China period, as well as the "second-generation ethnic policy" represented by "abolishing the autonomous regions" and "abolishing ethnicities". Such views have induced fierce disputes in the academic circle and have directly affected the practice of ethnic matters by having a quite negative influence on the relationship among ethnic groups.

Against such background, we should reconfirm the "Three Confidences" by comparing China to other countries, including confidence in our ethnic policy. To this day, the west has yet to provide the "perfect solution" for ethnic problems. Historically, when dealing with issues of races, ethnics, natives and immigrants, western nations have, for a long period of time, resorted to measures such as slavery, discrimination, separation, deporting, forced assimilation or even genocide. So, theories and practices of racism, nationalism, colonialism, chauvinism and fascism are all from the west. Even after the abolition of slavery, they still stuck to systems and policies such as "segregated but equal" (USA), "neither equal nor segregated" (Brazil) and "apartheid" (South Africa). The white supremacy ideology is deep-seated.

It was till the 1960s, under the influence of the American civil rights movement, that the western nations, driven by the politics of recognition, finally started to practice the doctrine of cultural diversity with equal-rights policies. These "equal-rights policies", which favored ethnic minorities like negroes in matters such as education, employment and loans, were in fact very similar to the supportive measures in Chinese ethnic policies. We are not claiming that the western nations learned or borrowed from our ethnic policies but they were indeed behind China in realizing the benefit of such policies or measures when it comes to respecting differences and narrowing disparities. In other words, China is leading the world in solving ethnic problems. There is no need for us to be self-contempt, thinking we're inferior to the west in everything.

Certainly, as for implementation of these policies and measures, there is room for progress on the effect of our systems, laws and policies designed to solve ethnic issues. It's directly related to the fact that China is still in the primary stage of socialism. Progressiveness of theories, ideologies and system design are part of the superstructure, which must be built on a decent economic foundation. This is why the Communist Party and the state keep stressing the key role of economic development in solving ethnic issues. That is to say, we should build a firm material foundation if we are to take full advantage of our political system. Deng Xiaoping once pointed out that "poverty is not socialism". The essential characteristic of socialism is common prosperity, which means, without common prosperity, real equality among all ethnic groups is impossible, ergo, advantages of socialism and its basic political systems like autonomy of ethnic regions won't pan out, either. Therefore, when comparing China to other countries, we can't ignore the most prominent of our national conditions: that China is in the primary stage of socialism. Nor can we so optimistically or even blindly believe that developed countries in the west have already solved ethnic issues in general (races, ethnics, religions, languages, immigrants, etc.).

It's undeniable that China has some intractable ethnic problems at hand, among which separatism is the most imminent. Causes of these problems include both "historical legacies" from when China was divided by imperialist powers and actual influence of shifting international political situation, among which the disintegration of the Soviet Union and the dramatic transformation of Eastern Europe were the most critical. Based on those events, the western nations have made many predictions such as "the triumph of nationalism over communism", or "all multinational socialist states will disintegrate after the USSR", etc. These predictions, obviously, were targeting China. And it was this very intention of using ethnic issues to tear China apart that, simultaneously (1990), gave rise to "Taiwanese independence", "Tibetan independence" and "East Turkestan".

In fact, western nations have never extricated themselves from the predicament of ethnic issues. When the socialist camp following the "Soviet Pattern" disintegrated, the western world, from a more macroscopic view, put forth the "End of History" theory, arguing that the development of capitalism may signal the end-point of humanity's social evolution. However, the so-called Third Wave of Nationalization incited by the west had caused conflicts in many countries and regions, among which conflicts of racial, ethnic and religious origins were the most prominent. Plus, these conflicts mostly happened in developing capitalist countries, and some in the developed world, thus giving birth to the theory, which understates ideology and social system, known as the "Clash of Civilizations", a hypothesis that Confucian, Islamic and Christian civilizations will have "fault line" conflicts against each other.

The west tried to transform the Islamic world in the name of eliminating "clash of civilizations" and they carried that strategy out through wars, oppressions, humiliations and contempt. The repercussions, in return, have substantially reshaped the western world. Starting with 911, the terror went on, with the Danish comic incident, burning of the Koran and the attack on Charlie Hebdo. Evidently, the west is facing crisis with issues such as race, ethnic, religion and immigrants. Those problems didn't go away in highly-developed societies. There have been multiple cases of separatism in ethnic regions, too. Federalization of Belgium, Quebec's status as "a nation within Canada", pursuit of independence in Basque and Catalonia, the Scottish Independence referendum, those are hardly new topics. There is institutional crisis in developed countries of the west.

Since 1970, the US, Canada and Australia started practicing policies favoring cultural diversity, a trend then affected Europe and many developing capitalist nations. Today, these policies have fallen into the so-called "predicament of recognition." Successively in 2010 and 2011, the German Chancellor and the British PM officially announced the failure of cultural diversity policy. The 2014 European Human Rights Report issued by the European Commission stated that democracy, rule of law and human rights conditions in Europe today are going through an unprecedented crisis that is the worst since the end of Cold War. Europe is faced with human rights conditions including racism and racial discrimination, corruption, human-trafficking and indulgence of hate speech. Discrimination against minorities existed in 39 European nations, police brutality in 23, exclusion and discrimination against Romani people in 23. In many European countries, poverty and high unemployment rate have aided the growth of conflicts and extremism. Over the past year, the streak of police brutality against black people have induced waves of protests in America, which came as a revelation to Americans that racism is far from being overcome in their country, and that racial issues remain unsolved.

All those examples only prove that unlike the old saying, stones from other hills won't necessarily serve to polish jade in this one. To quote President Xi Jinping: There is no such thing as a divine intervention when it comes to political systems. We can't simply conclude that we are missing something and want to borrow that just because other countries have it and we don't. Or that something must be a burden that we need to get rid of just because we have it but other countries don't. Both are cases of oversimplified and one-side thinking, which are plainly incorrect.

We do need to borrow helpful results from foreign political civilizations, but never shall we abandon the very root of Chinese political system. With a land area of 9.6 million square kilometers and a population consisting of 56 ethnic groups, where on earth will China find an existing model to follow? And who can tell us what to do? Therefore, when tackling or solving ethnic issues, we must do our due diligence even if we did draw on the experience of others, so we can have an accurate comprehension of their national conditions and the basis of their policies.

There have been propositions that, while criticizing and calling for removal of "ethnicity" from the Chinese ID card, suggest we should draw on the "Singaporean experience", which refers to a society where all ethnic groups get along peacefully and blend with each other in a shared pride of their Singaporean identity, especially the "Public Housing Plan" that provides a common residence for all ethnic groups. Now let's try and figure out how this "Singaporean experience" works. Singapore is a multiracial city state. The problems caused by race are not called ethnic issues but racial issues. The "Public Housing Plan", which is part of Singapore's "Ethnic Integration Policy", started with the "race" label on the Singaporean ID card. The communities were built according to respective proportion of the three major races. Now the Singaporean ID card bears symbols of over 90 races. For example, when a Chinese descendant marries a Malay, the parents are entitled to choosing the race for their children when applying for birth certificates. They can be Chinese, Malay Chinese or Chinese Malay. With the population of immigrants grows in Singapore, races are increasing too. And the government simply puts more labels on the ID card. Every citizen is entitled to a once-in-a-lifetime right to change their racial identity decided by their parents. If they chose not to change it the ID will stick with them for the rest of their lives. Whether you want to buy a house or enjoy any sort of race-related policies, your ID is the only proof.

Then, what is wrong with our ID card? Would we feel more Chinese had we remove ethnicity from the card? We must not resort to such reversed thinking, let alone blindly borrowing from others whatever experience we deemed good for solving ethnic issues or enhancing solidarity. We must understand what 's behind those experiences. Singapore does have these policies and they do put labels on the ID card, in a far more complicated and specific fashion than ours. Of course, the race factor doesn't just exist on the Singapore ID card. The three major races dominate the entire society in all aspects. Road signs are printed in English, Chinese, Hindu and Malay; Food is served according to their preferences and taboos. Companies must hire, by a certain proportion, people of different races. So, the Singaporean identity, of state or nation, is based on recognition of racial diversity.

In that sense, China's ethnic policies are made out of respect for the history and are in accordance with our national conditions, as well as complying to the public opinion and reflecting the fundamental principle of unification. Back when the People's Republic was newly founded, we began to identify each ethnic group in an effort to show recognition and respect for the multinational status of our country. This was way before the Singaporean policies, which came out after violent racial conflicts. Some may believe that by removing ethnicity from the ID card we can somehow consolidate identification of the Chinese nation or even eliminate ethnic issues for good as in the "Singaporean experience". Not only is that a naïve self-fooling illusion but it's far from the actual situation of Singaporean. Moreover, by simply removing ethnicity from the ID card, can we really get rid of discriminative regulations, which have been repeatedly criticized by the central government, such as special security check or inspection upon check-in against individuals of certain ethnic groups? Or, had we remove ethnicity from the ID card, is it still necessary for us to promote exchange, communication, intermingling and shared living space among all ethnic groups? What path and pattern a country follows in solving ethnic issues are the result of the combined influence of factors such as national conditions, history, economic/social situation and cultural traditions. As we stress the importance of the Three Confidences, the first task is to insist on solving ethic issues on our own terms.

Stay firmly on the correct path of tackling ethnic issues with solutions of Chinese characteristics. Explore and innovate. Those are significant political principles elucidated by Xi Jinping himself at the 2014 Central Ethnic Work Conference. This path, which is based on respect for history and is in accordance with our national conditions as well as complying to the public opinion, is a great achievement made by generations of CPC members through their practice of tackling ethnic issues. It is an achievement that we must cherish. Even if our goal isn't fully achieved in anyway, it is only because the economic foundation supporting the advantages of our political system hasn't reached an ideal level. So, we must have a proper view of our national conditions. We shall not feel complacent because of our achievements, nor shall we think we're inferior and feel self-contempt or even blindly follow others because the advantages of our political system haven't panned out due to inadequate support from economical/social development.

Domestic Ethnic Policy Is an Advantage When Pursuing the Belt and Road Initiative

The Belt and Road Initiative can be summarized as a master plan for China's all-directional opening-up, which includes "the Road, the Belt, the Corridors and the Bridge." The Corridors refer to three corridors, namely China-India-Bangladesh, China-Pakistan and China-Mongolia-Russia. And the Bridge is the continental bridge connecting Europe and Asia. This master plan of opening-up is rooted in China's neighboring areas and, by establishing a free trade zone among countries along the Belt and Road, aims to accelerate construction of infrastructure and realize connectivity. The ultimate goal is to achieve a development of mutual-benefit and shared success with tighter cooperation, more convenient exchanges and more integrated interests.

In 2010, the China-ASEAN Free Trade Zone was officially established. It is the most populous free trade zone in the world, with a population of nearly two billion, a total trade volume of six trillion and a total GDP of 9 trillion USD. On top of that, China has signed free trade agreements with Singapore, Pakistan, New Zealand, Chile, Peru and Costa Rica. Another five free trade zones are under negotiation respectively with the Gulf Cooperation Council, Australia, Norway, Iceland and Switzerland. Substantive talks on China-Australia and China-South Korea free trade zones have ended and await official signing.

Accelerating implementation of the free trade zone strategy is an important part in a new round of China's opening-up, which aims to build a new system of open economy and to gain the upper hand in economic development and international competition by opening up actively. Supported by neighboring areas, a high-standard global network of free trade zones is to be formed. Entering the new century, world oil gas resources have been shifting gradually to Middle East, Central Asia, Russia and the Americas in terms of supply and export, with the US, Europe and Asia-Pacific being the largest consumers. Those countries and regions, such as Middle East, Central Asia, Russia, Asia-Pacific and Europe, overlap with some of those along the Energy Silk Road, which is part of the Belt and Road.

As is known to all, energy is both the driving force and the bottleneck for development. China has the third largest energy production in the world but its per-capita amount is far below the world average. China's per-capita proven reserves are only as 33% of the world average. Since 1992, the gap between production and consumption in China has been growing increasingly. The imbalance between energy consumption and supply is becoming more and more obvious. According a projection by the IEA, 68% of China's total oil demand will rely on export by 2020. China is the world's largest importer of coals. And as shown by EIA's data, in 2012 China had overtaken the US as the largest net importer of oil in the world. Currently, over 60% of China's oil import comes from turbulent Middle East and North Africa. And most of that oil comes to China through centralized transportation at sea. Four fifth of all crude oil transportation goes through the Strait of Malacca so China is faced with the Malacca Dilemma, which hampers China's energy security. The open development of the Belt and Road Initiative aims to achieve "barrier-free" mutually-beneficial cooperation and shared prosperity between China and the world.

Some 2000 years ago, industrious and brave people of Eurasia had explored and discovered multiple routes of trade and cultural exchange among civilizations in Asia, Europe and African. The routes were then called by a joint name, the Silk Road. For thousands of years, the Silk Road spirits, of peaceful cooperation, openness and inclusiveness, mutual-interest and shared prosperity, have been passed on for generations and greatly propelled the progress of human civilization. It was a significant connection that facilitated prosperity and development of countries along the routes as well as a symbol of exchange and cooperation between east and west. It is a historical and cultural legacy shared by the entire world.

As stated in *Vision and Action*, the initiative will enhance people-to-people and cultural exchanges, and mutual learning among the peoples of the relevant countries, and enable them to understand, trust and respect each other and live in harmony, peace and prosperity. The Initiative is harmonious and inclusive. It advocates tolerance among civilizations, respects the paths and modes of development chosen by different countries, and supports dialogues among different civilizations on the principles of seeking common ground while shelving differences and drawing on each other's strengths, so that all countries can coexist in peace for common prosperity.

People-to-people bond provides the public support for implementing the Initiative. We should carry forward the spirit of friendly cooperation of the Silk Road by promoting extensive cultural and academic exchanges, personnel exchanges and cooperation, media cooperation, youth and women exchanges and volunteer services, so as to win public support for deepening bilateral and multilateral cooperation. All efforts to enhance people-to-people and cultural exchanges, and mutual learning among the peoples of the relevant countries, and enable them to understand, trust and respect each other and live in harmony, peace and prosperity are important cultural measures demonstrating a country's soft power. We can negotiate and build roads and reach agreements but what really matters is the people to people bond.

People to people bond is one of many crucial propellers for the building of the Belt and Road as well as the domestic and diplomatic condition for two-way opening-up. The most powerful and effective bond between peoples is culture and culture includes language. How can we establish a people to people bond if we can't even speak the same language? President Xi Jinping stressed at last year's Central Ethnic Work Conference that the most critical measure for solving ethnic problems and handling ethnic affairs is solidarity and the most useful one is having support of the people. Will of the people is the ultimate politics. Solidarity is all about sincere and honest communication. Language is a crucial link in people to people bond. Without speaking the same language people can't communicate so they can't understand and therefore can't reach agreements. This is a very important and practical issue for ethnic regions in two-way opening-up.

People to people bond is the principle of practice for solving ethnic problems with Chinese solutions. It reflects characteristic of those solutions such as respect for history, compliance with national conditions and public opinion. We must devise the opening-up of the Belt and Road and the Corridors in accordance with our ethnic policies. Ethnic minorities on the border have, for historical reasons, a traditional relationship with neighboring countries and regions. And that relationship is consisted of bonds through language, culture, customs and religion.

For example, owing to its diversity in ethnicity, culture, language and religion, Xinjiang is well-connected with nations in central and west Asia, even with others in the Muslim world. Is this an advantage or disadvantage for China? Ethnic minorities who inherit those cultures should play a bigger role in establishing people to people bond during the opening-up process in the building of the Belt and Road.

Xinjiang abounds with fruits

Opening-up of the Belt and Road will induce unprecedented transformation of position for ethnic minorities on the border. Meanwhile, we're faced with challenges in regional economic and social development, especially development and progress of all ethnic groups. As the core area of two-way opening-up, we need to reach out to the world and invite them in whilst dealing with infiltration and impact of hostile forces and the Three Evils. This requires us to stay firmly on the right path of solving ethnic problems with Chinese solutions and at first fully implement, domestically, the advanced ideas of China's ethnic policies.

Meanwhile, we'll cultivate a team of industrial workers and other talents from ethnic groups on the border, especially minorities so they can become developers and builders, as well as practitioners of two-way opening-up. This is a critical advantage of people to people bond. China's ethnic policies are highlighted by the respect for differences and efforts to narrow disparities. Implementation of the party's ethnic policies is the fundamental guarantee for accelerating economic and social development of ethnic regions on the border and the foundation for the advanced ideas of China's ethnic policies to spread outward through the building of the Belt and Road.

We often say that diplomacy is the extension of domestic affairs. Ideas and principles of China's ethic policies, which are part of our domestic affairs, can be extended to the practice of opening-up as well. A good example would be the practice of Chinese petroleum companies in Central Asia. For instance, in Turkmenistan, Chinese petroleum companies operated in a highly localized fashion with a 3–7 ratio of Chinese and local employees. They also made a great effort training managers and workers. Not only do Chinese companies show great respect for local culture, customs and habits, but they also learn the language and make contributions to public welfare such as livelihood issues. And that's how people to people bond can generate economic and social benefits. These practices are the very essence of China's ethnic polices, which are part of our domestic affairs. And these successful examples go to prove the effectiveness and validity of China's ethnic policies. Meanwhile, this requires our ethnic polices to be better and more thoroughly implemented domestically, especially in border regions inhabited by ethnic minorities so that the success of our domestic policies will have a positive influence on the Belt and Road endeavors. In retrospect, we must ask ourselves the following questions. How localized were our state-own enterprises when exploring resources in ethic regions on the border? Have they encouraged employees to learn the language or respect the culture of ethnic minorities? Have they spent time and money to train and recruit local peasants and herders as employees of state-own enterprises? If diplomacy is the extension of domestic affairs then ethnic policies, which are part of our domestic affairs, shouldn't be another case of "export rejects".

In response to a question raised by a certain Islamic country when Chinese authorities were introducing the Belt and Road vision to foreign ambassadors and corporations, one of them said, as we tried to explain the ideas of partnership and mutual-learning, that "China will obtain some experience in associating with Muslim people from this cooperation." That sounded unbelievable. In China, there are ten ethnic groups who have Islamic faith and the total Muslim population is over 20 million, which is not a small figure. The ten ethnic minority groups are members of the Chinese nation. Out intertwined destiny has stood the test of time in a course of millenniums. Couldn't that have given us "some experience in associating with Muslim people"? Ethnic diversity is an advantage in our development just as diversity in religion is also an advantage in China's opening-up. Grasp of ethnic and religious policies is of great importance in the opening-up of the Belt and Road. It's not that we lack experience in this field but we should extend our experience in ethnic solidarity and harmonious co-existence of religions to our opening-up practice.

President Xi Jinping once pointed out, by quoting Sima Qian's comment on the history of Pre-Qin, Qin and Han dynasties, that "Uprisings were normally started in the Southeast but fruits were gained by those in the Northwest." and as for building the Belt and Road, that "it is great news for ethnic regions, especially those on the border. We must thoroughly carry out the western development strategy and accelerate the pace of opening-up and development of border regions so as to explore new space for our country's development." This "new space" includes the advantage of cultural diversity of ethnic regions on the border as well as the advantage of people to people bond in people-oriented development pursued by all ethnic groups who inherit that cultural diversity. In that sense, the basic principles of respecting differences and narrowing disparities in China's ethnic policies are completely compatible with the humanism and spirit of mutual benefit advocated by the Belt and Road Initiative.

In conclusion, by building the Belt and Road, China aims to build a community of shared-interest that is interconnected, open and inclusive. The core spirits of this community are wide consultation, joint contribution and shared benefit instead of seeking dominance in reginal affairs or geopolitics. Countries along the routes are not pieces but chess players. Diplomacy is the extension of domestic affairs, effectiveness of which will determine success or failure of the opening-up. We will firmly and persistently implement decisions made by the Central Ethnic Work Conference, and carry forward building of the Belt and Road, as part of the "Four Comprehensives" strategy. Cultural diversity is the most important soft power for border regions and basic principles of China's ethnic policies are crucial advantages in the opening-up.

Shiyuan Hao Assistant to the President, Chinese Academy of Social Sciences; Secretary General, Presidium of Faculty CASS; Member the 11th and 12th CPPCC Recipient of the State Department's Special Government Allowances; Winner National Outstanding Middle-aged and Young Experts Awards Foreign Academician Mongolian Academy of Sciences.

Position and Role of Yunnan in China's Westward Opening-up Strategy

Jianhua Fan

The Belt and Road Initiative is an inevitable choice as Chinese economy transforms towards "new normal" and a critical strategic plan as China carries out the all-directional opening up westward. After 30 years of reforms and opening-up, China's economic achievements have attracted worldwide attention. However, the three-decade-long rapid growth, at the cost of environment while supported by benefits of policy and demographic dividend, has worn out its momentum. In the economic new normal, how China can breach the barrier and secure a future of heathy, ordered and sustainable development is a fundamental issue of great realistic and historical significance. Standing at a historical turning point and on the right course set by the 18th CPC National Congress, the people of China look towards the future while considering the reality as they made the critical choice to carry out the all-directional opening-up campaign. Through strategic moves such as the Belt and Road Initiative, China's macro economy will enter the international market full-scale as well as actively participate in international cooperation and competition. Chinese economy will propel and contribute to world economy in globalization while promoting healthy development of domestic economy so as to achieve mutual benefit as well as worldwide peace and development. In light of this strategy, president Xi Jinping, during his visit to Yunnan in 2015, defined Yunnan as: a province that will forwardly participate and serve national strategies by establishing itself as a bridge connecting China to south and southeast Asia. This definition pointed Yunnan in the right direction and gave the province an established goal and position.

J. Fan (✉)
Yunnan University, Kunming, China

W. Liu (ed.), *China's Belt and Road Initiatives*,
https://doi.org/10.1007/978-981-13-0101-8_12

The Silk Road and China's Westward Opening-up Strategy

Some 2000 years ago, the ancient Silk Road greatly facilitated economic and cultural exchanges between east and west as well as played a positive role in the thriving of the Han Dynasty. The Northern Silk Road was a main route connecting China to the west whereas the Southern Silk Road (the Sichuan-India Route) was the very first route that connected China to the outside world. The Maritime Silk Road was an international artery connecting China, via the South China Sea, to the Pacific and the Indian Ocean as well as the five continents of the world.

Throughout history, the center of the world has been moving from one place to another: Asia (during the Yuan Dynasty)—Europe (pre-WW2)—North America (post-WW2). This continuous shift was in fact a transfer of power and was determined by geopolitics, the center of which moves constantly. Just as described by Alvin Toffler in *Powershift,* that as one power falls another rises. In economic globalization, uneven growth of economy will give rise to a revolution of wealth. And as the superstructure depends on the economic foundation, politics must explore and find larger space for economic expansion. The Belt and Road framework, with focuses on the New Silk Road and westward opening-up, is a critical strategic decision in China's pursuit of all-directional opening-up as well as economic and cultural connectivity. It will push the Eurasian economic community forward into deeper integration. A new geopolitical setup will once again have a profound influence on the world. Implementation of the Belt and Road Initiative, the AIIB and the Silk Road Fund, as proposed by president Xi Jinping, is a scientific decision in the "new normal" of world economy.

From perspectives of geopolitics and regional economic development, reactivating the Southern Silk Road is an inevitable choice for China's Indian Ocean strategy as well as a practical need for building the China-India-Myanmar economic corridor. Trade and exchanges between Southeast Asia, South Asia, Europe and China's central, southwest, even the entire western area as a whole would be much more economical and convenient going through southwest China and the Indo-China Peninsula to the Indian Ocean. The Southern Silk Road strategy has, on some level, changed the structure of regional economy. For example, southwest China, northeast India, Myanmar and Bangladesh are relatively underdeveloped and cooperation was rather limited on the provincial level. To lift the building of "economic corridor" to the state level will, via the four countries, propel joint development of the three most important regions in Asia. By taking advantages of the four nations' geographic adjacency and economic complementariness as well as existing cooperation and enormous potentials, further economic cooperation will be carried out and an economic belt of strong appeal and great influence will be established. With the pace of integration of regional economy fastened, it is imperative for the four nations to harness their geopolitical advantages and consolidate political mutual-trust as well as further investment and trade, promote connectivity and enhance cultural exchanges.

Rebuilding the 21st Century Maritime Silk Road is a practical choice for the Pacific Rim to intensify international cooperation and preserve regional security as well as promote common development. China's economic strength can't thrive without maritime trade. Ever since the ancient times, the Maritime Silk Road has, in a spirit of peaceful trade, facilitated economic development through the opening-up of foreign trade. Throughout its history of maritime exploration, China had never resorted to military threats, violent conquering or barbaric occupation in its endeavors. Instead, China, supported by its mighty comprehensive strength, chose to promote free trade, peaceful exchanges and friendly associations. Over the past 2000 years, with the exception of the Yuan Dynasty when its army, after having taken Eurasia by storm, fiercely invaded countries and regions such as Vietnam, Java and Japan, China during other dynasties has managed to maintain a peaceful relationship with neighboring countries off its shores, especially when it comes to maritime trade. Ant that was, among other things, the reason why the Maritime Silk Road was praised as "the road of friendship" and "the road of peace". In terms of economic exchanges, the Maritime Silk Road greatly promoted the opening-up policies since the Tang and Song dynasties. In the millennium between Western Han and Yuan, Chinese dynasties had always encouraged maritime trade in a spirit of opening-up. The Song Dynasty, which was the only regime that didn't carry out commerce-restraining policies, was the most important maritime trader in the world then.

Back then China had the largest fleets of junks and merchant ships in the world and they were making frequent trips to Arabia, East Africa, India, Southeast Asia, Japan and Korea. Historically, the Maritime Silk Road has made a significant contribution to friendly exchanges between China and the world as well as promoting social, economic and cultural prosperity of all countries. Given current development, for the Pacific Rim to enhance international cooperation and preserving regional security as well as promoting common development, the most practical choices would be the following: rebuilding the 21st Century Maritime Silk Road, innovating mechanisms for regional cooperation, eliminating barriers for integration of regional economy and investment facilitation, fastening implementation of the strategy on free trade zones, expanding cooperation on trade and investment, establishing a new pattern of integrated regional economy.

Strategic Significance of Yunnan's in Terms of Geography and History

What position and role does Yunnan possess in the Belt and Road endeavors? Yunnan is the frontier and window in China's southwestward opening-up. The province has an obvious geographical advantage being linked to domestic regions such as Tibet, Sichuan, Guizhou and Guangxi as well as to foreign countries such as Myanmar, Laos and Vietnam. With an over-4000-km long borderline, which accounts for nearly one fifth of China's entire land border, Yunnan is the most efficient route connecting China to Southeast Asia and South Asia. Its unique advantage lies in its role of reaching the Pacific and the Indian Ocean as well as connecting East Asia, Southeast Asia and South Asia. As the frontier and window of China's southwestward opening-up, Yunnan, just like its neighboring areas in Myanmar, Laos and Vietnam, is an ethnic region on the border. And such areas are marked by geographic linkage, multiple routes, same ethnic origins and language, as well as close ties with peoples on the other side of the border. Meanwhile, Yunnan is blessed with a harmonious community of all ethnic groups, amicable relationship with neighboring countries, lasting solidarity among different peoples as well as stability on the border. The province also possesses abundant mineral, water and biological resources as well as resources of ethnic culture and tourism therefore has great potential for development. These unique geographic, historical and cultural conditions have determined Yunnan's tight relationship with neighboring countries in terms of ethnicity as well as ethnic issues.

Gate of Myanmar on China-Myanmar border

Yunnan has a special position for Chinas when it comes to macro geo-economics. In the new situation with intricate development of globalization, incessantly-intensified international and regional economic cooperation as well as enhanced implementation of China's Western Development strategy, Yunnan is going through rapid social and economic development. The urban agglomeration in central Yunnan is generating an even stronger driving force and the region is more capable of independent development. The whole province has achieved notable economic and social development thus bears the fundamental conditions to fasten the pace of development and become a regional hub for exchanges with South Asia and Southeast Asia. The strategic importance of Yunnan's geographic location is reflected in its being the connecting point between the two largest economic circles and belts in China. Eastward, Yunnan is linked with the Pearl River Delta and the

Yangtze River Delta; Southward, it can reach Hanoi, Bangkok, Singapore and Rangoon through eastern, central and western routes of the Trans-Asian Railway that is currently under construction; Northward, it's linked to Sichuan and China's inland; Westward, through Myanmar, it can reach Chittagong, Bangladesh and, via the Indian Ocean, enter the South Asia subcontinent and then Middle East to reach Malatya, Turkey and then enter Europe in the northwest or Arica in the southeast. Yunnan will serve as a crucial hub and platform for opening-up in southwest and central China. Cooperation-wise, it is the connection between southwest China and southeast Asia and South Asia. Therefore, domestically, Yunnan can reach Guangdong and Guangxi through Sichuan and Chongqing as well as the Yangtze River Economic Belt or it can reach the Maritime Silk Road through the Pearl River Delta, Hong Kong and Macau. Internationally, Yunnan can serve as a window in the southwestward opening-up and establishes itself as an international artery linking China with Southeast Asia and South Asia thus further improves quality and level of China's openness along the border. An all-directional opening-up pattern is to be formed. By enhancing mutually-beneficial cooperation with neighboring countries, China will promote common development, further friendly relationship among neighbors, facilitate coordinated development of the region, lift border and remote areas out of poverty, carry forward better and faster development in the southwestern zone of the Silk Road Economic Belt and achieved common prosperity among all ethnic groups as well as harmony and stability along the border.

Yunnan has been considerably important in China's diplomatic history too. The Southern Silk Road in southwest China was an ancient route that, stretching through Sichuan and Yunnan, connected Myanmar and India before reaching Southeast Asia, South Asia, West Asia and Europe. The ancient Southern Silk Road was the earliest recorded outward route in Chinese history. During the Pre-Qin period, it was already an important passage between China and the west with Yunnan on the earliest frontier and starting point of international communication. Yunnan is at the geographic center of Asia and all destinations in Asia can be reached from Kunming by plane within five hours. The province's unique geographic advantage lies in its accessibility in all directions such as Sichuan, Chongqing and Central China in the north, Vietnam, Laos, Thailand and Cambodia in the south, Myanmar, India and Pakistan in the west. It's located on the intersection between the north-south international artery and the third east-west Eurasian continental bridge starting from Shenzhen. As the frontier in China's opening-up towards Southeast Asia, South Asia, Middle East, South Europe and Africa, Yunnan is of great importance both historically and geographically. Pinpointing Yunnan's economic coordinates from a global perspective and we'll find that it is the pivot point linking East Asia, Southeast Asia and South Asia as well as the artery of the six great rivers of Asia. Each of those rivers, namely Dulong (Irrawaddy River), Nu (Salween River), Lancang (Mekong River), Jinsha (Yangtze River), Yuan (Red River) and Nanpan (Pearl River), can go through a naturally-formed international route in Yunnan. This brand-new perspective has put Yunnan on the irreplaceable role as the pivot connecting East Asia, Southeast Asia and South Asia.

Yunnan was an international transportation artery and supply route during World War Two. The Stilwell Road, built on the base of former Southern Silk Road, along with the Hump Course, which followed the same path as the Southern Silk Road, were the only international transportation arteries as well as supply routes for the rear area during the Anti-Japanese War. The Stilwell Road stretched from Lego, a border town in Northeast India, through the port of Houqiao and Tengchong to Longling. Finally, the two lines met and linked with the Burma Road to create a new international route during wartime. Being the most convenient route connecting the two most populous and fast-developing countries in the world, the Stilwell Road is the most efficient and economically attractive land route for China to enter the South Asian market so it bears both economic and historical significance. As a key point on the international route and a geographic pivot of strategic importance, Yunnan is entrusted with new missions in terms of international diplomacy, geopolitics as well as regional economic and cultural endeavors.

By establishing Yunnan as a hub in China's opening-up towards South Asia and Southeast Asia, the Central Government has determined Yunnan's role, which is a crucial move in China's effort to coordinate domestic development and all-directional opening-up as well as intensify implementation of the Western Development strategy, promote coordinated regional development and long-term stability along the border. It is also a rare historical opportunity for Yunnan to further accelerate its development to achieve goals compatible with the scientific outlook. Yunnan's role, as determined by the central government, decides that Yunnan is to lead the coordination of domestic development and opening-up. Supported by urban economic circle in central Yunnan, a geographical layout, equipped with the opening-up economic belt along the border and a domestic economic corridor, is to be formed. Focusing on building of the economic cooperation zone on the border and the cross-border economic cooperation zone, cross-border transportation and infrastructure such as ports and trunk highways are to be perfected. Trade and logistics as well as cross-border tourism are to be cultivated and developed with the stress on export-oriented processing and manufacturing industry. The overall goal is to form an economic belt along the border. Therefore, as has been determined, Yunnan's responsibilities and missions are in expanding and enhancing land communication with Southeast Asia and South Asia as well as in constructing the domestic section of the international artery from Yunnan to Southeast Asia and South Asia.

The Trans-Asian Railway and the Kunming-Bangkok Road between Yunnan and the ASEAN are both important pivots connecting China to Southeast Asia and South Asia. Construction of transportation infrastructure such as railway, highway, waterway and airway in Yunnan Province has accomplished preliminary goals. Other infrastructure such as energy, water conservancy and information facilities are fairly adequate too. In recent years, there has been sufficient development in economic cooperation between Yunnan and countries in Southeast Asia and South Asia. Cooperation in trade, investment and technology, in particular, has stepped up to a new level. The China-ASEAN Free Trade Zone has been officially established. Cooperation in the Greater Mekong River Sub-region has been intensified and

Regional cooperation among Bangladesh, China, India and Myanmar has been incessantly growing. So have Pan-Pearl River Delta Regional Cooperation and cooperation between Yunnan and Shanghai, Guangdong, Hong Kong, respectively. Transportation corridors, represented by the Trans-Asian Railway and the Kunming-Bangkok Road, have effectively facilitated economic and cultural exchanges. The consistent exploration in accelerating opening-up of territory and space in Yunnan and the ASEAN has transformed Yunnan from being on the edge of opening-up to the front.

Yunnan is the actual undertaker as China enhances building of the corridor of economic cooperation among Bangladesh, China, India and Myanmar as well as cooperation in the Greater Mekong River Sub-region. In perspective of traditional international politics, China is located on the edge of the Eurasian Continent and its coastal line is on the edge of the edge. With America's "pivot to Asia" and China's "pivot to the Pacific", which is called by some foreign press a countermeasure against the American strategy, plus the "problematic system" resulted from territorial disputes over East and South China Sea, China is under multiplied strategic pressure on the eastern front thus is in desperate need of a new strategic rear area that has ample space for the release of enormous strategic endogenous force. In that sense, the geopolitical setup of the "China-India-Myanmar-Bangladesh Corridor" is extremely important. Economically, the Corridor extends regional connectivity to South Asia and has played a crucial role in exerting Yunnan's geographic advantages, further consolidating political mutual-trust, intensifying investment and trade as well as promoting connectivity and cultural exchanges. Meanwhile, Yunnan's participation in the cooperation in the Greater Mekong River Subregion has brought about opportunities for development of railway, highway and water transport, as well as economic cooperation, and foreign trade, which greatly boosted growth of consumer demand. Directly transporting, at Chittagong, Bangladesh, oil and gas of Middle East into Yunnan through the land route is the most efficient choice that is economical and of great importance in China's oil strategy.

Yunnan Should Be the Core Hub that Connects the Three Roads, Two Oceans and Three Continents

Yunnan is the connecting point among the Three Roads (Northern, Southern and Maritime Silk Road). Economically speaking, the province's opening-up along the border is based its value of economic geography in national strategy. As the hub connecting China to Southeast Asia and South Asia, Yunnan will play an important role as the pivot in the opening up along the border, which is part of China's all-directional opening-up. In terms of strategic value, Yunnan will reflect the combined significance of national strategy, regional development along the border, China-ASEAN international cooperation and international sub-regional cooperation between China's border regions and Southeast Asia/South Asia. Specifically, by going north to Chengdu through Kunming, west Yunnan can be directly linked to the Northern Silk Road. Or, by merging with the Northern Silk Road through the ancient tea route and the Tang-Tibet ancient route, an international artery connecting Yunnan, along with Guangdong and Guangxi, to eastern Europe is to be formed so as to facilitate the economic integration of southeast/southwest China and the world.

Yunnan is the pivot connecting the west and the south of China in the mutually-complementary opening-up. Given the geographic pattern of the Maritime Silk Road, Yunnan is where its eastern route meets the southern one. Based in Yunnan, South China Sea can be reached through Vietnam's Ho Chi Minh or Haiphong and then the Indian Ocean through the Strait of Malacca. Meanwhile, entering India's inland is possible through north Myanmar so Europe can be reached by going through Calcutta and Bombay to the Indian Ocean and then the Red Sea. Thus, Southern Silk Road and Southern Maritime Silk Road can be linked so as to effectively connect the west and the south of China in mutually-complementary opening-up. It will serve to break the administration-based division and, geographically, facilitate the balancing of population, economy, resources and environment. Eventually, a pattern of coordinated regional development will be achieved with complementary advantages, mutual-facilitation and shared prosperity between the west and the south of China.

Yunnan is an important gateway connecting mid-west China to the world. Ever since the ancient times, open border has always been Yunnan's characteristic and advantage as well as potential. Yunnan is the most convenient land route connecting China to Southeast Asia and South Asia. What makes the province unique is that the Pacific and the Indian Ocean can be reached through it and that East Asia, Southeast Asia and South Asia can be connected. Since the State Department approved the national border economic cooperation zone status of Ruili, Wanding and Hekou in 1992, Yunnan has made the most of this platform of economic cooperation along the border and intensified the implementation of the strategy focusing on industry, trade and services as it incessantly promoting exploration and opening-up till the province became an important window where western and central regions of inland China meet the world through opening-up along the border. By putting the focus on along-border and cross-border economic cooperation zones, Yunnan has completed the construction of infrastructure such as cross-border transportation, ports and border routes. And this has greatly facilitated China's national strategy of fitting the economic development of central and western regions into the globalization framework. As an important outward window for central regions of inland China such as Guizhou, Hunan and Hubei as well as southwestern regions like Chongqing, Sichuan and Guangxi, Yunnan, on one hand, has the support from said provinces and, on the other hand, plays the role of the frontier of opening-up for them. Therefore, Yunnan's importance as the doorway in China's all-directional opening-up, especially the westward opening-up in the Belt and Road Initiative, has been deemed prominent.

By using Yunnan as the pivot, the Northern Silk Road can be linked to the Maritime Silk Road through the Pearl River Economic Belt and the Yangtze River Economic Belt. At the same time, the Pacific and the Indian Ocean can be connected to form a network linking China with Asia, Europe and Africa.

Basic Conditions and Developmental Responsibilities of Yunnan

Building of the New Silk Road Economic Belt can be carried out through integrated transportation passageways while basing on infrastructure along the route and key cities. Trade and production within the region are to be optimized to promote regional economic integration and, eventually, common economic and social development of the region. Trade and investment facilitation, economic and technological cooperation, as well as establishing free trade zones, are the trilogy in building the New Silk Road Economic Belt.

China's effort in building a friendly and trustworthy relationship with all its neighboring countries has created a favorable international environment for Yunnan's extensive participation in regional economic cooperation. With the Western Development strategy further implemented, Yunnan is to play a bigger role in nationwide coordinated regional development. This has largely boosted Yunnan's independent development capability and competitiveness in the region. Launch and development of the Bangladesh-China-India-Myanmar Economic Corridor and Greater Mekong Subzone Cooperation, as well as approving and advancing of Ruili National Key Experimental Zone For Development and Opening-up, National Border Economic Cooperative Zones and Honghe Comprehensive Bonded Zone are of great importance serving as examples for expanding international exchanges and cooperation, elevating foreign trade, propelling all-directional, wide-ranging and multi-dimensional opening-up. Meanwhile, it will greatly push forward the strengthening of international integrated transportation passageways, international shipping and logistics systems, port facilities as well as construction of urban infrastructure.

Construction of high-grade highways has been completed in domestic sections of international arteries connecting Yunnan and neighboring countries such as Vietnam, Laos, Thailand, Myanmar and India, along with seven trunk highways to neighboring provinces. There is a total of 16 international cargo routes, which serve as arteries of Yunnan's opening-up, leading to Laos and Vietnam among other countries in the Greater Mekong Subzone. As for railways, the network consisting of eight routes entering, and four exiting, Yunnan has been included in national planning. Currently, four new routes, namely Yunnan-Tibet, Kunming-Shanghai, Chongqing-Kunming and Yunnan-Guangxi, are under construction. Also in preparation are railways linking China to Vietnam, Thailand and Myanmar as well as to South Asia through Myanmar. In the aviation front, construction of 12 airports over the entire province has been finished. Four airports, which are under construction, will be operational by 2015. A total of 393 courses have been established, flying to and from 130 cities. As for water transport, Yunnan has been developing the Lancang River-Mekong River international shipping route and constructing the golden waterway of Yangtze River as well as the You River-Pearl River water passage. Tourism and water transport of reservoirs and lakes are being developed too. The province has had some impressive accomplishments with its water transport construction.

General connectivity, of roads, electricity, postal services, commerce, oil, gas, trade and customs, between Yunnan and neighboring countries has been fully intensified, providing a firm foundation for the westward opening-up. With closer economic ties with other countries and gradual realization of Yunnan's transportation master plan, the province will lift itself completely out of remoteness, isolation and underdevelopment in terms of transportation. Yunnan, as well as the entire southwest China, will become the bridge and link among the three major economic belts in the world—China's eastern coast, Southeast Asia and South Asia. Competition among and integration of the three economic forces will make Yunnan a strategic high ground for China's regional cooperation endeavors, leading to energetic regional economic prosperity in China's Western Development. With goals such as connecting China and the world through land, rivers and seas (the Pacific and the Indian Ocean) as well as the effort to established connectivity of roads, electricity, postal services, oil, gas, currencies and customs, Yunnan has made substantial breakthroughs in building the international artery opening up southwestward. Orderly connection between China and Southeast Asia/South Asia in transportation, trade, communication and energy has been gradually realized. The mutually-beneficial strategic ties among regions are consolidated, laying a firm foundation for Yunnan to fully implement the westward opening-up strategy.

In recent years, by taking over industries transferred from eastern China, Yunnan has built a base for said industries. Also, Yunnan has played an active role in promoting regional cooperation in the Pan-Pearl River Delta as well as regional innovation and all-directional opening-up and development. After having successfully held the 3rd China-South Asia Expo and the 23rd Session of China Import & Export Fair in Kunming, Yunnan has become an important platform for exchanges and cooperation between China and Southeast Asia/South Asia. Meanwhile, through strengthening cooperation with neighboring provinces, cities and regions such as Shanghai, Zhejiang and the Pan-Pearl River Delta, Yunnan has won strong domestic support in the all-directional opening-up strategy.

By high-standards and in a pertinent fashion, Yunnan takes over transferred industries form coastal regions and overseas in the effort to attract, with its various economic zones at all levels and along the border as well as at ports and in areas under special supervision from the customhouse, investment and loans from home and abroad. By building the transferred-industries base that, supported by the greater southwest and linked with the Pearl River Delta and the Yangtze River Delta, opens to the rest of China as well as South Asia and Southeast Asia, Yunnan will, through opening-up and cooperation, have strong support domestically in the westward opening-up.

Yunnan should try and "export" its culture while becoming a hub connecting China with South Asia and Southeast Asia.

The Southern Silk Road is not only a route of trade but also one of culture. Trough the South Silk Road, Chinese culture transmitted westward and enriched the civilizations in South Asia, Central Asia, West Asia as well as Europe and the Mediterranean Region, making a significant contribution to the development of ancient civilization of the world. We should consider the possibility of combining resources along tourist routes in the province into a quality route of cultural tourism while promoting cooperation among tourism zones along the route. Meanwhile, Yunnan should join hands with Sichuan and Guizhou in discovering, protecting and studying ancient cultures such as Ba & Shu, Dian, Yelang as well as Buddhism, Bimo, Dongba and other splendid minority cultures along the ancient route so as to lay the foundation for the joint declaration (with Myanmar, Thailand and India) of the Southern Silk Road as a World Heritage.

Cross-border ethnic groups in Yunnan are crucial carriers and intermediaries for the "export" of Chinese culture. Cultural similarity between cross-border ethnic groups is the reason for direct or indirect influences on China and its neighboring countries in aspects of politics, economy and diplomacy. And Yunnan has one of the largest populations of cross-border minorities in China. There are 17 ethnic groups, namely Han, Zhuang, Dai, Buyi, Yi, Hani, Lagu, Lisu, Jinpo, Achang, Nu, Wa, Dulong, De'ang, Bulang, Miao and Yao, living cross-border with Myanmar, Thailand, Vietnam and Laos. Each ethnic group can live cross-border in different number of countries and the same ethnic group may be known by other names in different countries, but people of every single group were closely connected historically and have a strong sense of ethnic identity in reality. They share similar customs and habits and speak the same language. Ever-lasting traditions, which became critical factors in economic and cultural exchanges, include visiting relatives and friends, intermarriage and trade, sharing farmland and pastures, religious activities and festive events.

Great efforts should be made to promote exchanges and cooperation of cross-border cultures as well as import/export of cultural products and transmission of those cultures. Not only should we advance cultural exchanges on the government level, but we should also enhance exchanges on civic levels. And those exchanges should include both public and commercial sectors. We should have an overall view on cultural issues and further our studies as well as fully discover and exert resources of ethnicities, cultures and talents to form a powerful combine force of cross-border cultural transmission.

Epilogue

Yunnan's unique geographic location as well as traditional connections and strategic role are of great significance both historically and realistically in China's all-directional opening-up, in which the province will play a special part, especially in the westward opening-up strategy focusing on the building of the Belt and Road. As far as national strategy is concerned, the Central Government should put a new premium on Yunnan in the Belt and Road opening-up master plan and entrust it with important historical responsibilities. In China's all-directional westward opening-up, Yunnan and Xinjiang should be the two windows and hubs that are of the same importance. As for Yunnan itself, we should strive for the position of one of the key provinces in China's westward opening-up strategy and benefit from relevant supporting policies, which will help Yunnan get transfer payment for critical national projects as well as understanding and trust from provinces, cities and regions in central and western China, so we can really play our role as the bridge and hub and window in serving the grand strategy of national all-directional opening-up.

Jianhua Fan President Yunnan Federation of Social Science Circles; Researcher, Doctoral Supervisor; Former Director Institute of Ethnic Literature, Yunnan Academy of Social Sciences, Deputy Director General, Department of Culture of Yunnan Province; Deputy Director, Yunnan Cultural Industry Office; Member Committee of Experts CIC; Member State-own Cultural Assets Management Committee, Ministry of Finance; Guest Professor, Chinese Academy of Governance, Yunnan University, Communication University of China, Consultant of Cultural Industry Development, Qinghai and Gansu Province.

On Further Promoting Cooperation and Development of Education Among Countries Along the Belt and Road

Zhenyuan Qu

In September and October 2013, president Xi Jinping, during his visits to central and southeast Asian countries, successively put forth the initiatives of building the Silk Road Economic Belt and the 21st Century Maritime Silk Road (hereafter referred to as the Belt and Road). The initiatives immediately received utmost attention from the international community. On March 8th, 2015, three ministries under the State Department co-issued *Vision and Actions on Jointly Building Silk Road Economic Belt and 21st-Century Maritime Silk Road* so as to accelerate building of the Belt and Road, which, for the purpose of peace and common development of the world, aims to promote economic prosperity and regional cooperation as well as exchanges and mutual learning among countries along the routes through coordinated economic policies, free flow of production essential elements, efficient allocation of resources and deep integration of markets.

In this systematic project of the century, the role of public education, especially higher education in China, lies in the following endeavors: actively promoting cooperation and development of education among countries along the Belt and Road, cultivating massive number of talents, fit into and lead the building of the Belt and Road while improving and transforming into a powerhouse of higher education. All these issues require deliberate consideration, thorough research, bold exploration and persistent practice.

Z. Qu (✉)
Ministry of Education, Beijing, China

W. Liu (ed.), *China's Belt and Road Initiatives*,
https://doi.org/10.1007/978-981-13-0101-8_13

Building the Belt and Road Is a National Strategy of Great Importance

The Belt and Road Initiative, focusing on economy and trade as well as connectivity, is a great vision of multinational strategic cooperation for the purpose of mutual-benefit and shared prosperity. It inherits the ancient Silk Road and improves it by connecting the Asian-Pacific Economic Circle and Europe through a total 64 countries in Central Asia, Southeast Asia, South Asia, West Asia and East Africa.[1] It is an open, inclusive, and balanced initiative for economic cooperation excluding no country. It is neither an entity nor operating on a closed system so all countries and economies are welcomed to join in for common and cooperative development. The Belt and Road strategy breaks the old mode of regional development, which was scattered and loose, and establishes a new pattern. Those along the Belt and Road are mostly emerging economies and developing countries, with a total population of 4.4 billion and a combined economic magnitude of 21 trillion USD, respectively accounting for 63 and 29% of the world's population and economy. Most of these countries have a growing economy and a myriad of resources so there is great potential and much space for cooperation.

Building of the Belt and Road comprises three aspects, namely economy and trade, regional order and cultural exchange, that focus on policy, infrastructure, trade, financial and people-to-people connectivity, the specific contents of which are as follows. Policy: promote intergovernmental cooperation and jointly provide policy support for the implementation of practical cooperation and large-scale projects; Infrastructure: construction of infrastructure of transportation, energy, communication and technical standard systems; Trade: free trade zones, customs cooperation, certification and accreditation, standard measurement, statistical information, cross-border e-commerce, modern service trade, agriculture, forestry, animal husbandry and fisheries, cross-border economic cooperation zones; Financial: bilateral currency swap and settlement, the bond market in Asia, the Asian Infrastructure Investment Bank, BRICS New Development Bank, Shanghai Cooperation Organization financing institution, the Silk Road Fund, financial regulation; People-to-people: cultural and academic exchanges, personnel exchanges and cooperation, media cooperation, cooperation in jointly running schools, training of professionals, holding culture years, cooperation in tourism, medical assistance, sports exchanges, cooperation in science and technology, vocational skill training, social security management, public administration management, legislative bodies, major political parties and political organizations, cities, think tanks, nongovernmental organizations, poverty reduction, protection of World Cultural Heritages, biodiversity and ecological protection. Building the Belt and Road

[1]According to up-to-date statistics, there are totally 64 countries along the routes, which include 43 in Asia, 16 in Eastern Europe, four in Commonwealth of Independent States, and one in Africa. Source: China Social Sciences Network, http://www.cssn.cn/gj/gj_gwshkx/gjjj/20150408_1578096.shtml, April 8th, 2015.

comes with wide space for cooperation on agriculture, textile industry, chemical industry, energy, transportation, communication, finance, science and technology, medical industry, etc. *Vision and Actions* also indicates that building the Belt and Road will, through further expanding the opening-up, make huge efforts to establish the following projects deeply integrated into globalization: one strategic stronghold, two cores, seven cooperative zones, nine inland open economic high grounds, 15 coastal ports and two international hub airports, which cover construction of infrastructure and industrial development in most provinces, cities and regions. Implementation of this strategic vision will certainly have a significant, profound and lasting influence on China in terms of sustainable development, social solidarity and stability as well as technological and cultural exchanges.

The Belt and Road are to be built, in a cooperative and win-win spirit, for common development and shared prosperity. It is a path, of peace and friendship, leading to better understanding and wider exchanges. It is a strategic, long-term, high-level and all-directional grand strategy that bears great purposes and noble missions such as comprehensive opening-up, coordinated development and rejuvenation of the Chinese nation. Achieving these goals require massive space and prolonged timespan so it'll take more than a few years to accomplish. We should focus on several points-in-time such as 2020, 2030 and 2050 as we, through much contemplation during the historical process of realizing the great dream of rejuvenating the Chinese nation, determine goals and priorities at different stages and then carry out the plans in a logical and organized fashion. We should keep in mind at all times the chronicity, difficulty and complexity of these endeavors. Neither indifference nor impatience will get us anywhere. Over the past 20 years or so, emerging economies represented by the BRICS have been growing more than twice as faster as developed industrialized countries. This trend has transformed international politics and economy just as the Belt and Road will.

Building of the Belt and Road Raises New Missions and Demands for the Opening-Up of Chinese Education

"A community of common destiny for mankind" is a new concept, which has been repeatedly stressed by the Chinese government in recent years, about human development. The 18th NCCPC put forth for the first time that "we should raise awareness about human beings sharing a community of common destiny. A country should accommodate the legitimate concerns of others when pursuing its own interests; and it should promote common development of all countries when advancing its own development." President Xi Jinping frequently mentioned "the community of common destiny for mankind" at over 60 different occasions, especially diplomatic events. In April 2013, Tsinghua University launched the Schwarzman Scholars program, in honor of the occasion president XI Jinping himself issued a letter that offered a further explanation of the idea of "a community of common destiny for mankind". He stated in the letter that "today's world is a community of common destiny comprised of all countries. To overcome challenges in human development, all peoples need to join hands and make collective efforts." The Belt and Road Initiative focuses on building a community of common interest, destiny and responsibilities through all-directional and practical cooperation. In a word, proposing and implementing the Belt and Road Initiative are China's practice to fulfill the promise of building a community of common destiny for mankind.

Education Undertakes a Unique Mission

One shall maintain his own integrity while in poverty and benefits the world while in prosperity. Such has been the sense of responsibility in Chinese culture since ancient times. And the same belief is reflected today by the Belt and Road Initiative and the establishment of AIIB as China approaches its goal of becoming a well-off society in an all-around way and then a moderately prosperous one. Most countries along the Belt and Road had suffered from exploitation and suppression of colonial times. Restrained by historical and some of their own conditions, many of those countries are still struggling with poverty, hunger and turmoil. And as they move towards modernization they're faced with a world whose political and economic structures were formed upon the law of the jungle since World War Two. They can't possibly follow the same path of the developed countries when it comes to strategy and mode of development so they are in desperate need of an international system that is more just and reasonable as they seek development in a world advancing into multi-polarization economic globalization, cultural diversity and social informatization.

Historically, the modern international system is formed through three stages. The first stage was marked by the Peace of Westphalia signed in 1648, which established the principles pertaining to sovereignty, territory and independence, ergo, lay the foundation for legitimate status nation states in international affairs. The second stage was when sovereign states formed alliances represented by the 1815 Congress of Vienna, during which European powers, through negotiation, ended the warfare that lasted for over 20 years and re-divided the continent. As a result, alliances such as NATO, ASEAN or SCO, which are formed by different nations, became important measures in dealing with international affairs. The third stage was when the United Nations system, founded after WW2 by the dominant US and other countries, came to rise. Organizations such as the UN itself, World Bank and the IMF have been playing major parts in world peace as well as economic and social development since after WW2. However, we should see that those international organizations are the reflection of the world back then in terms of political and economic power. The organizations were established on values and ideas of western nations and possess decisive powers that were never formed upon justice and fairness so they are outcomes of history and in a certain sense limited. For example, any major decision of the IMF requires at least 85% of all votes to pass while the US holds 17.65% of votes thus has the right to veto. Leadership of these organizations are recommended, routinely, only by Europeans and Americans. Since the last international financial crisis, requests for increasing shares of votes of China, Russia and India in World Bank and the IMF have been raised for several times and have all been denied at the US Congress. Therefore, China should, from an objective angle of human development and imbalance of global development, step up to a higher ground and take on its responsibilities as a developing major power. Those responsibilities include insisting peaceful development, making efforts to build a harmonious world, playing a constructive role, preserving its image as a responsible major power, participating in current international system while devoting itself to reforming the system in a responsible manner to make it more just and reasonable. By proposing the Belt and Road Initiative and establishing the AIIB, China is, in an effort to benefit the world, promoting the mutually-beneficial and win-win integration of Eurasia as well as the ideas of community of common interest and destiny.

When speaking of "a community of common destiny", president Xi Jinping especially stresses, as far as education is concerned, that "education should go with the trend and, through frequent interaction and exchanges, promote awareness of all human knowledge and cultures as well as recognition of all nations' struggles and visions in order to facilitate mutual-understanding among students of all countries. By building up an international perspective and inspiring innovations, education can endow people with great ambitions so they're determined to contribute their wisdom and strength to peace and development of mankind. Admitting the "trend" is a pre-condition for recognizing the mission that comes from that trend, which education should fit itself into and support.

Since the founding of the People's Republic, especially after the reforms and opening-up, Chinese education system has been going with the trend and gradually formed a pattern of opening-up that is all-directional, multi-dimensional and wide-ranging, thus established China's status as the world's largest source of overseas students and the no. 1 study destination in Asia. After joining the WTO, China, among other major powers, has established a relatively higher level of commitment in opening-up of education. In some respects, China's opening-up level surpasses some developed countries and many key reference countries that haven't committed to opening-up of education. Our most imminent task, under the new circumstances, is to comply with the new trend and seize the opportunity to take on new missions and demands in building the Belt and Road.

Top Priority of Education, Especially Higher Education, Is Talent Training

What kind of talents does it take to build the Belt and Road? The initiative's massive number of projects can be broken into three categories: One, infrastructure of transportation, information and energy, trade and investment, energy resources, connectivity of currency and finance, all these above fall into the category of engineering and economy; Two, regional eco-environment protection, maritime cooperation, policy connectivity, these are in the category of regional politics and order; Three, regional language and culture, technology and humanities, connectivity of health and tourism, these are cultural exchanges and cooperation. Contents of these strategies include infrastructure, technology, capital, currency, trade, culture, policy, ethnics, religion, etc. Every single one of those needs talents cultivated through education, especially higher education.

First of all, construction of infrastructure needs professionals of engineering, project design and management in various fields. According to an evaluation report of Asian Development Bank, during 2010–2020, Asian countries will need to invest a total of 7.97 trillion USD in construction and maintenance of infrastructure, which involves 989 transportation projects and 88 of cross-border energy projects.[2] Building these projects needs hundreds of thousands, if not millions, of talents of engineering, design and construction, quality control and guarantee, economic management, in industries such as railway, pipelines, electricity, roads, ports and communication. It also requires collaboration of experts in fields such as engineering, politics, economics and management.

Second of all, with many companies setting up offices overseas, talents who speak local languages and are familiar with local politics, economy, culture, customs and geography are in great demand, especially ones who have a grasp of politics, economy and local customs of countries in Southeast Asia, South Asia, Central Asia, Northeast Asia, even West Asia. There are 64 countries along the routes of the Belt and Road and more are joining in. On the other hand, talents of minority languages are in severe shortage so it's almost a crisis. Education and talent training for those languages are imperative. Plus, the Chinese public think English is the only useful foreign language and developed countries the only foreign countries. These notions are certainly incompatible with the fact that China is promoting diversity, opening-up and international perspective among its people therefore must be corrected.

Thirdly, regional economic exchanges and forming of order need a great many of foreign trade professionals. The Belt and Road is shaping up to be the third trade axis in the world other than the ones in the Atlantic and the Pacific. Currently, countries along the Belt and Road have a combined GDP of 20 trillion USD (approximately 1/3 of the world total) and economic growth in countries of the region relies largely on cross-border trade. In 2000, the average degree of dependence on foreign trade among those countries was 32.6%, which increased to 33.9% by 2010 and 34.5% by 2012, considerably higher than world average (24.3%) of the same period. Calculated with World Bank's data, during 1990–2013 direct investment in global cross-border trade was growing by 7.8 and 9.7% annually

[2]Higher Education in China Must Set a New Course in Response to the Belt and Road Strategy, Source: www.gmw.cn Guangming Daily, May 12th, 2015.

whereas the numbers with countries along the Belt and Road were 13.1 and 16.5%. During 2010–2013, after the financial crisis, foreign trade and net influx of foreign capital in countries along the Belt and Road grew by 13.9 and 6.2% respectively, which were 4.6 and 3.4% points higher than world average. In the next ten years the total export volume of countries along the Belt and Road is expected to increase to about 1/3 of world total.[3] The first project after the founding of AIIB was building of the Silk Road Economic Belt, which demands a huge number of talents who understand capital operation, currency circulation, rule-making in trade as well as international regulations.

Enhance Education on Understanding Nations and Cultures, Promote People-to-People Bond

People-to-people bond is, essentially, intermingling of cultures. Culture is a universal spiritual product of the entire human race. Mutual-benefit is the guarantee for successful implementation of the Belt and Road strategy whereas people-to-people bond is the social foundation. Generally speaking, countries along the routes all have complicated national conditions, religious beliefs, geopolitics and public opinions as well as political turmoil that is unpredictable and huge disparities among areas, classes and religious sects. Only by thoroughly understanding the people's demands and will as well as eliminating misunderstanding and misjudgment can we advance cooperation. Only when scholars, entrepreneurs, governments, non-governmental organizations and the people fully comprehend their historical/cultural background and public opinions can we better carry out this strategy. That's what's missing now and that's why we need to enhance and further education on understanding nations and cultures. In order to effectively achieve policy connectivity, we need to provide, for decision-makers, ambassadors of the Belt and Road endeavors as well as government officials, entrepreneurs and civilians who engage in relevant practices, comprehensive and in-depth training of knowledge in fields such as history, geography, language, culture, religion and politics. And in order to establish people-to-people bond, we need to cultivate a great many of civilians and cultural ambassadors, who have excellent diplomatic skills, positive influence as well as reputation and are traveling frequently among different countries, that fit into the entire society through NGO volunteering, academic studies and cultural exchanges.

[3]The Belt and Road Leads the Way in China's Future Opening-up, Shanghai Securities News, January 14th, 2015.

Statue of confucius

Education, Especially Higher Education, Must Make Efforts to Provide Intellectual Support as Well as Make Contribution of Wisdom

As was proved throughout history, a country's economic and political status in the global setup isn't immutable. The world is in a constantly changing state. To claim new positions in the latest shift of global landscape, Asian-Pacific nations must comply with the trend of reginal and global cooperation. Leften Stavros Stavrianos, eminent historian and the author of A Global History, once said that "Given the same geographical factors, the key to progress for mankind is approachability among peoples. Nations who have the most opportunities to influence and be influenced by others are the ones that will most likely thrive. As a matter of fact, environment would force them to thrive too, because not only are they faced with opportunities to develop but also the pressure of being eliminated." The Belt and Road Initiative acknowledges the uniqueness of each country along the routes in terms of development while establish a system of common development with countries sharing the role of center and source. This common development system decides that the study of critical subjects facing the community of common destiny for mankind must be intensified and that solutions must be proposed.

First of all, we must dedicate our wisdom to the building of a harmonious region governing system. Countries along the routes of the Belt and Road are mostly emerging economies whose international influence keeps growing with their development. Nevertheless, unfair and unreasonable international political and economic order, as well as disparity of economic and social development among different countries and regions, still exist. The gap between the poor and the rich keeps enlarging while regional conflicts and violence won't go away. All these problems are what makes governing the countries along the routes difficult. Challenges and problems as such need to be tackled by the common development system built up along with the Belt and Road.

Second of all, education should contribute to sustainable development of the region and the world, which are faced with severe challenges brought by issues such as global climate change, energy shortage, water crisis, forest protection, desertification, biodiversity preservation, environmental pollution, prevention of infectious disease, public safety, rapid urbanization in emerging economies, population explosion, resource shortage, etc. We need to face these challenges together and solve them by collaborating with each other. The fact that existing international rules are, obviously or subtly, dominated by developed countries of the west determines that it is unrealistic for us to count on them to fairly uphold the benefit of the whole world, let alone emerging economies. Countries along the Belt and Road have more in common as to how developing countries can benefit from participating global governance as well as dealing with and solving mentioned issues. We should join hands as we devise and adjust strategies and countermeasures that will facilitate sustainable development of the region and the world, for which purpose Chinese higher education must play a bigger role.

Thirdly, education must contribute to offering practical solutions for realistic issues. Various realistic issues, ranging from macro to micro, cultural to social, policy to engineering, human resources to technological bottleneck, etc., will occur during the building of the Belt and Road. We need to carry out research on social development of countries in the region and along the routes, scientific evaluation of differences and conditions of the countries involved, economic and cultural exchanges, research on international commercial cooperation, investigation and cultivation of as well as research on talent demands. We will also carry out research on prospective, pertinent and precautionary policies while provide scientific evaluation, strategic reflection and advancing planning concerning future development of the Belt and Road in five, fifteen and fifty years. We will enhance cooperative researches among states, ministries and departments, regional governments, institutions of higher education, industries and trades while offering, according to demands of decision-making, policy suggestions that are professional, constructive and feasible.

Further Advancing Education Cooperation and Development Among Countries Along the Routes

Implementation of the Belt and Road strategy not only requires us to actively participate in economic development, regional cooperation and planning of countries along the routes, but also examine domestic higher education for the purpose of fitting it into development of the Belt and Road strategy. Meanwhile, we'll learn from the international higher education system and, in a vision of rebuilding the world order, devise strategic layout and action plans for Chinese higher education to take part in the building of the Belt and Road. For example, providing talents and intellectual support for the joint effort of countries along the routes to build the Belt and Road, as well as promote cooperation and exchanges of economy, culture and education among countries along the routes, furthermore, by sharing the fruits of the Belt and Road endeavors, achieve win-win cooperation. These are China's obligations as a responsible great power and necessary actions by the Chinese higher education system.

Expanding Number of International Students in China and Cultivating Much-Needed Overseas Talents

International student programs have become important instruments for a country to lift its global influence as well as expand the education market. Since the reforms and opening-up, China has put a new premium on international student recruitment. As a result, scale and quality of international student programs in China have improved at a steady pace. According to the *Investigation Report on International Student Programs in China 2014*, there was a total of 377,000 international students from 203 countries and regions studying in 775 institutions of higher education, research institutes and other facilities of 31 provinces, regions and municipalities. By number of students, China has become the third largest study destination globally, accounting for 8% of world total. However, in the international education market, China is still falling way behind the largest beneficiaries such as the US, Australia and the UK. As far as sources of students are concerned, those from Asian countries are the largest group among international students in China. There was a total of 2,254,000 of them in 2014, accounting for almost 60% of all international students in China. In recent years, there has been such a steep rise in number of students from Southeast Asia that they have become the dominant group. Eight of the top ten countries by number of students are countries along the routes of Belt and Road, namely South Korea, Thailand, Russia, Japan, Indonesia, India, Pakistan and Kazakhstan. Certainly, for students from countries along the routes, demands of studying in China will keep increasing with the advancing of the Belt and Road endeavors.

For a long time, China's international student programs have been focusing on a few developed countries. Some universities believe only by recruiting students from western nations can the school demonstrate the level and quality of internationalization of its education. But from the angles of supporting the Belt and Road strategy as well as long-term goals of education, Chinese education system must dedicate itself to building a community of common destiny for mankind and benefiting the entire humanity. That's why we must, under the guidance of China's national strategy, expand numbers of international students in China, optimize the structure of our international student program, continue to admit students from developed countries and make extra efforts to increase the number of students from countries along the Belt and Road. China is to further multiply number of candidates for the government scholarship with a tendency to favor countries along the routes. Increased candidacy will be given mainly to students from those countries and it is a timely help that we must provide.

Another important subject requiring our input is how we can meet the demands of individual students while putting emphasis on supporting the building of the Belt and Road as well as improving the quality of education for international student programs. Currently, in terms of demands of students from different parts of the world, those from developed countries prefer studies of language while those from underdeveloped countries prefer degree courses such as medicine and engineering. Subjects and majors pertaining to connectivity of infrastructure, trade and capital, which are of great importance cooperation-wise in building the Belt and Road, are mostly not optional in China's international student programs. To change that, we must, from a strategic perspective, carry out over planning of curriculum options for students from countries along the Belt and Road as well as combine advantageous resources to build up distinctive subjects and majors relevant to the Belt and Road strategy for the purpose of attracting students who will, after finishing their studies in China, put their learning into good use when return to their home countries.

The key to improving the quality of education in China's international student programs is providing excellent and sufficient courses. To that end, institutions of higher education should take full advantage of contemporary information technology and, by combining online and offline services, develop curriculum systems and degree courses, in various languages, that can facilitate the building of the Belt and Road. We should strive, with high-quality resources and services of education, to meet the demands of the Belt and Road endeavors while devising quality courses and majors for international students in China.

Cultivating Highly-Qualified Professionals Through Integration of Education and Industry as Well as Schools and Enterprises

Building of the Belt and Road is a massive systematic project, which can only be carried out in a vigorous, organized and effective fashion with the support of higher education. As far talent training is concerned, the classification system must be insisted. Not only are we training elites, young talents and future leaders who fully understand international rules and can take on national missions, but also a great number of highly-qualified professionals who are competent for the construction of Belt and Road infrastructure. We must distinguish the order of priority when building the Belt and Road. For example, preferentially assign professionals to or train them for the building of major passages, traffic arteries, trunk routes, key points and links. We should pursue shoulder-to-shoulder development of education and industries through integration of both sectors by supporting cooperative education between institutions of higher education and enterprises in "going global" industries such as high-speed rail and telecommunication. Currently, there is a total of over ten thousand Chinese enterprises in countries along the Belt and Road with only a few cases of cooperative education. Meanwhile, we must improve flexibility of training programs so they become a convenient option. The purpose would be for the programs to follow the projects wherever they go. Furthermore, the programs should be at the same level as, if not ahead of, major infrastructure projects of the Belt and Road. Programs should be carried out locally and integrate training of highly-qualified professionals with undergoing projects.

As for composition of human resources, countries along the routes have yet to become aging societies. In 2013, an average of 67.5% of the total population in countries along the routes were aged 15–64 and labor force accounted for over 70% of the total population in 21 of those countries so labor resources were rather adequate. At the same time, infrastructure in those countries were among the medium-low level globally. According to the statistics of infrastructure in World Bank's logistics performance index, countries along the Belt and Road scored an average of 2.7 points (as oppose to the global average of 2.91 pts) in 2013. Abundant resources of labor power in countries along the Belt and Road, along with infrastructure awaiting construction, provide a vast market for education service for China's powerful high-end manufacturing industry as well as rich experience acquired through nearly ten years of international cooperative education, especially higher professional education. In recent years, Ningbo Polytechnic setup the Benin International Training Center to train local employees hired by Chinese companies and to propel economic development of African countries. Guilin Tourism University has been training tourism professionals for ASEAN nations such as Indonesia and Brunei. The Belt and Road strategy should borrow from and promote these successful experiences.

In conclusion, coordinated practices of "inviting in" and "going out" should, on the talent front, serve as the fundamental path supporting the Belt and Road endeavors.

Selectively Establishing Overseas Universities and Education Bases in Countries Along the Routes

In recent years, quality of Chinese higher education has won worldwide recognition. In more than ten years of its course, the plan to create world-class universities had cultivated a myriad of top-notch innovative talents and world-class disciplines as well as numerous world-leading original products, playing a big role in improving China's comprehensive strength. In particular, Chinese higher engineering education, which is one of the three major fields in the higher education system, had as of 2006 started establishing a professional certification system that would be globally authentic and compatible with existing system of engineers. The Chinese system applied and was permitted to join the Washington Accord in June 2013, indicating that the quality guarantee system for China's engineering education was internationally recognized. Currently, there are, at undergraduate level, totally 15,733 engineering majors, which have basically covered all important projects as required by the building of the Belt and Road. Therefore, Chinese higher education has already had an excellent foundation for "going out" and setting up overseas universities and education bases in countries along the routes. Meanwhile, we have accumulated some referential experiences. Currently, Chinese universities already have a considerable existence overseas. Overseas schools of Chinese universities set up as approved by the Ministry of Education include Xiamen University Malaysia, Soochow University Laos, Yunnan University of Finance and Economics Bangkok Business School and Beijing Language and Culture University Tokyo College. Meanwhile, there are over 90 programs involving 14 countries and regions, mostly in Southeast Asia. We have established bilateral and multi-lateral education exchanges and cooperation with over 180 countries and regions while having signed agreements with 41 countries and regions for mutual-admittance of diplomas and degrees. Also, more than ten nations along the routes, including Pakistan, Kazakhstan and Jordan, have invited China to set up overseas education programs in their countries. Therefore, by increasing investment in education when building the Belt and Road, as well as developing programs designed for countries along the routes and supporting the establishing of overseas universities or other education facilities, not only can we cultivate talents for economic and social development in those countries, but also establish extensive connections in the development of the world while discovering and training middle-aged and young talents who understand and are friendly to China, so we can build up a favorable reputation on international grounds and expand China's global influence. Meanwhile, this would be a strategic move propelling Chinese education to go global as well as a self-assigned mission for China to create an amicable relationship with its neighboring countries and seek benefit for countries along the routes.

Broadening Our International Vision and Advancing Chinese Higher Education

To meet the demands of building the Belt and Road, we must improve the weak links of disciplines and majors. For example, we must accelerate the training of non-universal language speakers as well as intensify the study of language demands in building the Belt and Road. Meanwhile, we'll make special developmental plans for languages and enlarge the size of talent pool for strategic foreign languages while accelerating the training of high-end talents who not only master languages of countries along the routes but also understand their national conditions and cultures. Currently, there are totally over forty official languages spoken in countries along the Belt and Road whereas Chinese universities can only offer courses for 20 of them. Most universities in developed countries offer hundreds of language courses when there are less than 70 languages listed as undergraduate majors by China's education authorities. Currently, problems such as insufficient number of non-universal language programs, inappropriate arrangement of language majors as well as shortage of language talents for key countries and regions have become the bottleneck hampering the advance of the Belt and Road projects. So, it is imperative that we treat the training of language talents as part of our national strategy and make an all-out effort to that end. We must build up existing and experienced universities that specialize in foreign languages as well as ones that offer non-universal language courses, turning them into centers for language learning and bases for studies of cultures and nations while cultivating more talents of language and culture for the Belt and Road endeavors.

Implementation of the Belt and Road strategy prompts us to, from a wider international perspective, carry out an overall examination of the quality of Chinese higher education and improve it. Chinese higher education can emerge above countries along the Belt and road or even the rest of the world only when the core factors such as majors, courses, teaching, practice and faculty have reached the top level by international standards and that, academically, Chinese-trained students can match their counterparts of the same expertise and level in developed countries. A more direct competition of quality of education on the international arena is taking place when some Chinese universities are running schools overseas. Therefore, we must put an emphasis on the comprehensive improvement of the quality of education as well as broadening the global vision for Chinese higher education and building up all types of institutions. In an effort to meet the demands of the Belt and Road endeavors, we are to change the pattern of talent training and adjust the structure of programs as well as advance reforms of teaching methods and examine the quality of talent training from an international perspective.

We should also, through building the Belt and Road, seize the opportunity to adjust the structure of Chinese higher education system and strengthen government support for higher education in central and western China as well as solve the problem of disparity between eastern and western China in higher education.

Not only will those actions benefit coordinated development of higher education in different regions nationwide, but also contribute to building the Belt and Road.

We must enhance cooperation among universities in countries along the routes when advancing building of the Belt and Road by standing together to face common challenges such as studies of politics, economy, culture and security so China can play a bigger role in international education governance. In January 2015, Xi'an Jiaotong University officially founded the Collaborative Innovation Centre for Silk Road Economic Belt while putting forth the concept of the "Silk Road Academic Belt". Meanwhile, the University Alliance of the New Silk Road was founded for cooperative researches on innovation, elite cultivation, policy-making consulting, professional services and settlement of disputes, all serve to help building and development of the Belt and Road. By making those efforts, the university has greatly improved its academic competence as well as its global influence so it should be seen as an excellent example.

The visionary concept of the Belt and Road aims to unite countries in Asia, Europe and Africa into a community of common interest and mutual benefit as well as a community of common destiny and shared prosperity. The people yearn for a better life, the foundation of which is high-quality education. Implementation and advance of the Belt and Road strategy have entrusted Chinese higher education system with new missions and responsibilities. This will certainly propel our great cause to move forward upon a new historical starting point as we accelerate the transformation from quantity to quality of higher education and make our due contribution to peace and development of the world as well as the progress of human civilization.

Zhenyuan Qu President, China Association of Higher Education; Member, National Advisory Council on Education; Member, National Guidance Council on Education and Exams; Deputy Director, 2013–2017 Committee of Experts for Evaluation of Undergraduate Teaching in Regular Higher Educational Institutions; Member, Approval Board for Constitutions of Higher Educational Institutions, Ministry of Education; Representative, The 11th and 12th CPPCC.

Appendix A
Vision and Actions on Jointly Building Silk Road Economic Belt and 21st-Century Maritime Silk Road

Issued by the National Development and Reform Commission,
Ministry of Foreign Affairs, and
Ministry of Commerce of the People's Republic of China,
with State Council authorization

Preface

More than two millennia ago the diligent and courageous people of Eurasia explored and opened up several routes of trade and cultural exchanges that linked the major civilizations of Asia, Europe and Africa, collectively called the Silk Road by later generations. For thousands of years, the Silk Road Spirit—"peace and cooperation, openness and inclusiveness, mutual learning and mutual benefit"—has been passed from generation to generation, promoted the progress of human civilization, and contributed greatly to the prosperity and development of the countries along the Silk Road. Symbolizing communication and cooperation between the East and the West, the Silk Road Spirit is a historic and cultural heritage shared by all countries around the world.

In the 21st century, a new era marked by the theme of peace, development, cooperation and mutual benefit, it is all the more important for us to carry on the Silk Road Spirit in face of the weak recovery of the global economy, and complex international and regional situations.

When Chinese President Xi Jinping visited Central Asia and Southeast Asia in September and October of 2013, he raised the initiative of jointly building the Silk Road Economic Belt and the 21st-Century Maritime Silk Road (hereinafter referred to as the Belt and Road), which have attracted close attention from all over the world. At the China-ASEAN Expo in 2013, Chinese Premier Li Keqiang emphasized the need to build the Maritime Silk Road oriented towards ASEAN, and to create strategic propellers for hinterland development. Accelerating the building of the Belt and Road can help promote the economic prosperity of the countries along the Belt and Road and regional economic cooperation, strengthen exchanges and mutual learning between different civilizations, and promote world peace and development. It is a great undertaking that will benefit people around the world.

W. Liu (ed.), *China's Belt and Road Initiatives*,
https://doi.org/10.1007/978-981-13-0101-8

The Belt and Road Initiative is a systematic project, which should be jointly built through consultation to meet the interests of all, and efforts should be made to integrate the development strategies of the countries along the Belt and Road. The Chinese government has drafted and published the Vision and Actions on Jointly Building Silk Road Economic Belt and 21st-Century Maritime Silk Road to promote the implementation of the Initiative, instill vigor and vitality into the ancient Silk Road, connect Asian, European and African countries more closely and promote mutually beneficial cooperation to a new high and in new forms.

A.1 Background

Complex and profound changes are taking place in the world. The underlying impact of the international financial crisis keeps emerging; the world economy is recovering slowly, and global development is uneven; the international trade and investment landscape and rules for multilateral trade and investment are undergoing major adjustments; and countries still face big challenges to their development.

The initiative to jointly build the Belt and Road, embracing the trend towards a multipolar world, economic globalization, cultural diversity and greater IT application, is designed to uphold the global free trade regime and the open world economy in the spirit of open regional cooperation. It is aimed at promoting orderly and free flow of economic factors, highly efficient allocation of resources and deep integration of markets; encouraging the countries along the Belt and Road to achieve economic policy coordination and carry out broader and more in-depth regional cooperation of higher standards; and jointly creating an open, inclusive and balanced regional economic cooperation architecture that benefits all. Jointly building the Belt and Road is in the interests of the world community. Reflecting the common ideals and pursuit of human societies, it is a positive endeavor to seek new models of international cooperation and global governance, and will inject new positive energy into world peace and development.

The Belt and Road Initiative aims to promote the connectivity of Asian, European and African continents and their adjacent seas, establish and strengthen partnerships among the countries along the Belt and Road, set up all-dimensional, multi-tiered and composite connectivity networks, and realize diversified, independent, balanced and sustainable development in these countries. The connectivity projects of the Initiative will help align and coordinate the development strategies of the countries along the Belt and Road, tap market potential in this region, promote investment and consumption, create demands and job opportunities, enhance people-to-people and cultural exchanges, and mutual learning among the peoples of the relevant countries, and enable them to understand, trust and respect each other and live in harmony, peace and prosperity.

China's economy is closely connected with the world economy. China will stay committed to the basic policy of opening-up, build a new pattern of all-round opening-up, and integrate itself deeper into the world economic system. The

Initiative will enable China to further expand and deepen its opening-up, and to strengthen its mutually beneficial cooperation with countries in Asia, Europe and Africa and the rest of the world. China is committed to shouldering more responsibilities and obligations within its capabilities, and making greater contributions to the peace and development of mankind.

A.2 Principles

The Belt and Road Initiative is in line with the purposes and principles of the UN Charter. It upholds the Five Principles of Peaceful Coexistence: mutual respect for each other's sovereignty and territorial integrity, mutual non-aggression, mutual non-interference in each other's internal affairs, equality and mutual benefit, and peaceful coexistence.

The Initiative is open for cooperation. It covers, but is not limited to, the area of the ancient Silk Road. It is open to all countries, and international and regional organizations for engagement, so that the results of the concerted efforts will benefit wider areas.

The Initiative is harmonious and inclusive. It advocates tolerance among civilizations, respects the paths and modes of development chosen by different countries, and supports dialogues among different civilizations on the principles of seeking common ground while shelving differences and drawing on each other's strengths, so that all countries can coexist in peace for common prosperity.

The Initiative follows market operation. It will abide by market rules and international norms, give play to the decisive role of the market in resource allocation and the primary role of enterprises, and let the governments perform their due functions.

The Initiative seeks mutual benefit. It accommodates the interests and concerns of all parties involved, and seeks a conjunction of interests and the "biggest common denominator" for cooperation so as to give full play to the wisdom and creativity, strengths and potentials of all parties.

A.3 Framework

The Belt and Road Initiative is a way for win-win cooperation that promotes common development and prosperity and a road towards peace and friendship by enhancing mutual understanding and trust, and strengthening all-round exchanges. The Chinese government advocates peace and cooperation, openness and inclusiveness, mutual learning and mutual benefit. It promotes practical cooperation in all fields, and works to build a community of shared interests, destiny and responsibility featuring mutual political trust, economic integration and cultural inclusiveness.

The Belt and Road run through the continents of Asia, Europe and Africa, connecting the vibrant East Asia economic circle at one end and developed

European economic circle at the other, and encompassing countries with huge potential for economic development. The Silk Road Economic Belt focuses on bringing together China, Central Asia, Russia and Europe (the Baltic); linking China with the Persian Gulf and the Mediterranean Sea through Central Asia and West Asia; and connecting China with Southeast Asia, South Asia and the Indian Ocean. The 21st-Century Maritime Silk Road is designed to go from China's coast to Europe through the South China Sea and the Indian Ocean in one route, and from China's coast through the South China Sea to the South Pacific in the other.

On land, the Initiative will focus on jointly building a new Eurasian Land Bridge and developing China-Mongolia-Russia, China-Central Asia-West Asia and China-Indochina Peninsula economic corridors by taking advantage of international transport routes, relying on core cities along the Belt and Road and using key economic industrial parks as cooperation platforms. At sea, the Initiative will focus on jointly building smooth, secure and efficient transport routes connecting major sea ports along the Belt and Road. The China-Pakistan Economic Corridor and the Bangladesh-China-India-Myanmar Economic Corridor are closely related to the Belt and Road Initiative, and therefore require closer cooperation and greater progress.

The Initiative is an ambitious economic vision of the opening-up of and cooperation among the countries along the Belt and Road. Countries should work in concert and move towards the objectives of mutual benefit and common security. To be specific, they need to improve the region's infrastructure, and put in place a secure and efficient network of land, sea and air passages, lifting their connectivity to a higher level; further enhance trade and investment facilitation, establish a network of free trade areas that meet high standards, maintain closer economic ties, and deepen political trust; enhance cultural exchanges; encourage different civilizations to learn from each other and flourish together; and promote mutual understanding, peace and friendship among people of all countries.

A.4 Cooperation Priorities

Countries along the Belt and Road have their own resource advantages and their economies are mutually complementary. Therefore, there is a great potential and space for cooperation. They should promote policy coordination, facilities connectivity, unimpeded trade, financial integration and people-to-people bonds as their five major goals, and strengthen cooperation in the following key areas:

Policy coordination

Enhancing policy coordination is an important guarantee for implementing the Initiative. We should promote intergovernmental cooperation, build a multi-level intergovernmental macro policy exchange and communication mechanism, expand shared interests, enhance mutual political trust, and reach new cooperation consensus. Countries along the Belt and Road may fully coordinate their economic development strategies and policies, work out plans and measures for regional cooperation, negotiate to solve cooperation-related issues, and jointly provide policy support for the implementation of practical cooperation and large-scale projects.

Facilities connectivity

Facilities connectivity is a priority area for implementing the Initiative. On the basis of respecting each other's sovereignty and security concerns, countries along the Belt and Road should improve the connectivity of their infrastructure construction plans and technical standard systems, jointly push forward the construction of international trunk passageways, and form an infrastructure network connecting all sub-regions in Asia, and between Asia, Europe and Africa step by step. At the same time, efforts should be made to promote green and low-carbon infrastructure construction and operation management, taking into full account the impact of climate change on the construction.

With regard to transport infrastructure construction, we should focus on the key passageways, junctions and projects, and give priority to linking up unconnected road sections, removing transport bottlenecks, advancing road safety facilities and traffic management facilities and equipment, and improving road network connectivity. We should build a unified coordination mechanism for whole-course transportation, increase connectivity of customs clearance, reloading and multimodal transport between countries, and gradually formulate compatible and standard transport rules, so as to realize international transport facilitation. We should push forward port infrastructure construction, build smooth land-water transportation channels, and advance port cooperation; increase sea routes and the number of voyages, and enhance information technology cooperation in maritime logistics. We should expand and build platforms and mechanisms for comprehensive civil aviation cooperation, and quicken our pace in improving aviation infrastructure.

We should promote cooperation in the connectivity of energy infrastructure, work in concert to ensure the security of oil and gas pipelines and other transport routes, build cross-border power supply networks and power-transmission routes, and cooperate in regional power grid upgrading and transformation.

We should jointly advance the construction of cross-border optical cables and other communications trunk line networks, improve international communications connectivity, and create an Information Silk Road. We should build bilateral cross-border optical cable networks at a quicker pace, plan transcontinental submarine optical cable projects, and improve spatial (satellite) information passageways to expand information exchanges and cooperation.

Unimpeded trade

Investment and trade cooperation is a major task in building the Belt and Road. We should strive to improve investment and trade facilitation, and remove investment and trade barriers for the creation of a sound business environment within the region and in all related countries. We will discuss with countries and regions along the Belt and Road on opening free trade areas so as to unleash the potential for expanded cooperation.

Countries along the Belt and Road should enhance customs cooperation such as information exchange, mutual recognition of regulations, and mutual assistance in law enforcement; improve bilateral and multilateral cooperation in the fields of inspection and quarantine, certification and accreditation, standard measurement, and statistical information; and work to ensure that the WTO Trade Facilitation Agreement takes effect and is implemented. We should improve the customs clearance facilities of border ports, establish a "single-window" in border ports, reduce customs clearance costs, and improve customs clearance capability. We should increase cooperation in supply chain safety and convenience, improve the coordination of cross-border supervision procedures, promote online checking of inspection and quarantine certificates, and facilitate mutual recognition of Authorized Economic Operators. We should lower non-tariff barriers, jointly improve the transparency of technical trade measures, and enhance trade liberalization and facilitation.

We should expand trading areas, improve trade structure, explore new growth areas of trade, and promote trade balance. We should make innovations in our forms of trade, and develop cross-border e-commerce and other modern business models. A service trade support system should be set up to consolidate and expand conventional trade, and efforts to develop modern service trade should be strengthened. We should integrate investment and trade, and promote trade through investment.

We should speed up investment facilitation, eliminate investment barriers, and push forward negotiations on bilateral investment protection agreements and double taxation avoidance agreements to protect the lawful rights and interests of investors.

We should expand mutual investment areas, deepen cooperation in agriculture, forestry, animal husbandry and fisheries, agricultural machinery manufacturing and farm produce processing, and promote cooperation in marine-product farming, deep-sea fishing, aquatic product processing, seawater desalination, marine biopharmacy, ocean engineering technology, environmental protection industries, marine tourism and other fields. We should increase cooperation in the exploration and development of coal, oil, gas, metal minerals and other conventional energy sources; advance cooperation in hydropower, nuclear power, wind power, solar power and other clean, renewable energy sources; and promote cooperation in the processing and conversion of energy and resources at or near places where they are exploited, so as to create an integrated industrial chain of energy and resource cooperation. We should enhance cooperation in deep-processing technology, equipment and engineering services in the fields of energy and resources.

We should push forward cooperation in emerging industries. In accordance with the principles of mutual complementarity and mutual benefit, we should promote in-depth cooperation with other countries along the Belt and Road in new-generation information technology, biotechnology, new energy technology, new materials and other emerging industries, and establish entrepreneurial and investment cooperation mechanisms.

We should improve the division of labor and distribution of industrial chains by encouraging the entire industrial chain and related industries to develop in concert; establish R&D, production and marketing systems; and improve industrial supporting capacity and the overall competitiveness of regional industries. We should increase the openness of our service industry to each other to accelerate the development of regional service industries. We should explore a new mode of investment cooperation, working together to build all forms of industrial parks such as overseas economic and trade cooperation zones and cross-border economic cooperation zones, and promote industrial cluster development. We should promote ecological progress in conducting investment and trade, increase cooperation in conserving eco-environment, protecting biodiversity, and tackling climate change, and join hands to make the Silk Road an environment-friendly one.

We welcome companies from all countries to invest in China, and encourage Chinese enterprises to participate in infrastructure construction in other countries along the Belt and Road, and make industrial investments there. We support localized operation and management of Chinese companies to boost the local economy, increase local employment, improve local livelihood, and take social responsibilities in protecting local biodiversity and eco-environment.

Financial integration

Financial integration is an important underpinning for implementing the Belt and Road Initiative. We should deepen financial cooperation, and make more efforts in building a currency stability system, investment and financing system and credit information system in Asia. We should expand the scope and scale of bilateral currency swap and settlement with other countries along the Belt and Road, open and develop the bond market in Asia, make joint efforts to establish the Asian Infrastructure Investment Bank and BRICS New Development Bank, conduct negotiation among related parties on establishing Shanghai Cooperation Organization (SCO) financing institution, and set up and put into operation the Silk Road Fund as early as possible. We should strengthen practical cooperation of China-ASEAN Interbank Association and SCO Interbank Association, and carry out multilateral financial cooperation in the form of syndicated loans and bank credit. We will support the efforts of governments of the countries along the Belt and Road and their companies and financial institutions with good credit-rating to issue Renminbi bonds in China. Qualified Chinese financial institutions and companies are encouraged to issue bonds in both Renminbi and foreign currencies outside China, and use the funds thus collected in countries along the Belt and Road.

We should strengthen financial regulation cooperation, encourage the signing of MOUs on cooperation in bilateral financial regulation, and establish an efficient regulation coordination mechanism in the region. We should improve the system of risk response and crisis management, build a regional financial risk early-warning system, and create an exchange and cooperation mechanism of addressing cross-border risks and crisis. We should increase cross-border exchange and cooperation between credit investigation regulators, credit investigation institutions and credit rating institutions. We should give full play to the role of the Silk Road Fund and that of sovereign wealth funds of countries along the Belt and Road, and encourage commercial equity investment funds and private funds to participate in the construction of key projects of the Initiative.

People-to-people bond

People-to-people bond provides the public support for implementing the Initiative. We should carry forward the spirit of friendly cooperation of the Silk Road by promoting extensive cultural and academic exchanges, personnel exchanges and cooperation, media cooperation, youth and women exchanges and volunteer services, so as to win public support for deepening bilateral and multilateral cooperation.

We should send more students to each other's countries, and promote cooperation in jointly running schools. China provides 10,000 government scholarships to the countries along the Belt and Road every year. We should hold culture years, arts festivals, film festivals, TV weeks and book fairs in each other's countries; cooperate on the production and translation of fine films, radio and TV programs; and jointly apply for and protect World Cultural Heritage sites. We should also increase personnel exchange and cooperation between countries along the Belt and Road.

We should enhance cooperation in and expand the scale of tourism; hold tourism promotion weeks and publicity months in each other's countries; jointly create competitive international tourist routes and products with Silk Road features; and make it more convenient to apply for tourist visa in countries along the Belt and Road. We should push forward cooperation on the 21st-Century Maritime Silk Road cruise tourism program. We should carry out sports exchanges and support countries along the Belt and Road in their bid for hosting major international sports events.

We should strengthen cooperation with neighboring countries on epidemic information sharing, the exchange of prevention and treatment technologies and the training of medical professionals, and improve our capability to jointly address public health emergencies. We will provide medical assistance and emergency medical aid to relevant countries, and carry out practical cooperation in maternal and child health, disability rehabilitation, and major infectious diseases including AIDS, tuberculosis and malaria. We will also expand cooperation on traditional medicine.

We should increase our cooperation in science and technology, establish joint labs (or research centers), international technology transfer centers and maritime cooperation centers, promote sci-tech personnel exchanges, cooperate in tackling key sci-tech problems, and work together to improve sci-tech innovation capability.

We should integrate existing resources to expand and advance practical cooperation between countries along the Belt and Road on youth employment, entrepreneurship training, vocational skill development, social security management, public administration and management and in other areas of common interest.

We should give full play to the bridging role of communication between political parties and parliaments, and promote friendly exchanges between legislative bodies, major political parties and political organizations of countries along the Belt and Road. We should carry out exchanges and cooperation among cities, encourage major cities in these countries to become sister cities, focus on promoting practical cooperation, particularly cultural and people-to-people exchanges, and create more lively examples of cooperation. We welcome the think tanks in the countries along the Belt and Road to jointly conduct research and hold forums.

We should increase exchanges and cooperation between non-governmental organizations of countries along the Belt and Road, organize public interest activities concerning education, health care, poverty reduction, biodiversity and ecological protection for the benefit of the general public, and improve the production and living conditions of poverty-stricken areas along the Belt and Road. We should enhance international exchanges and cooperation on culture and media, and leverage the positive role of the Internet and new media tools to foster harmonious and friendly cultural environment and public opinion.

A.5 Cooperation Mechanisms

The world economic integration is accelerating and regional cooperation is on the upswing. China will take full advantage of the existing bilateral and multilateral cooperation mechanisms to push forward the building of the Belt and Road and to promote the development of regional cooperation.

We should strengthen bilateral cooperation, and promote comprehensive development of bilateral relations through multi-level and multi-channel communication and consultation. We should encourage the signing of cooperation MOUs or plans, and develop a number of bilateral cooperation pilot projects. We should establish and improve bilateral joint working mechanisms, and draw up implementation plans and roadmaps for advancing the Belt and Road Initiative. In addition, we should give full play to the existing bilateral mechanisms such as joint committee, mixed committee, coordinating committee, steering committee and management committee to coordinate and promote the implementation of cooperation projects.

We should enhance the role of multilateral cooperation mechanisms, make full use of existing mechanisms such as the Shanghai Cooperation Organization (SCO), ASEAN Plus China (10 + 1), Asia-Pacific Economic Cooperation (APEC), Asia-Europe Meeting (ASEM), Asia Cooperation Dialogue (ACD), Conference on Interaction and Confidence-Building Measures in Asia (CICA), China-Arab States Cooperation Forum (CASCF), China-Gulf Cooperation Council Strategic Dialogue, Greater Mekong Sub-region (GMS) Economic Cooperation, and Central Asia Regional Economic Cooperation (CAREC) to strengthen communication with relevant countries, and attract more countries and regions to participate in the Belt and Road Initiative.

We should continue to encourage the constructive role of the international forums and exhibitions at regional and sub-regional levels hosted by countries along the Belt and Road, as well as such platforms as Boao Forum for Asia, China-ASEAN Expo, China-Eurasia Expo, Euro-Asia Economic Forum, China International Fair for Investment and Trade, China-South Asia Expo, China-Arab States Expo, Western China International Fair, China-Russia Expo, and Qianhai Cooperation Forum. We should support the local authorities and general public of countries along the Belt and Road to explore the historical and cultural heritage of the Belt and Road, jointly hold investment, trade and cultural exchange activities, and ensure the success of the Silk Road (Dunhuang) International Culture Expo, Silk Road International Film Festival and Silk Road International Book Fair. We propose to set up an international summit forum on the Belt and Road Initiative.

A.6 China's Regions in Pursuing Opening-up

In advancing the Belt and Road Initiative, China will fully leverage the comparative advantages of its various regions, adopt a proactive strategy of further opening-up, strengthen interaction and cooperation among the eastern, western and central regions, and comprehensively improve the openness of the Chinese economy.

Northwestern and northeastern regions. We should make good use of Xinjiang's geographic advantages and its role as a window of westward opening-up to deepen communication and cooperation with Central, South and West Asian countries, make it a key transportation, trade, logistics, culture, science and education center, and a core area on the Silk Road Economic Belt. We should give full scope to the economic and cultural strengths of Shaanxi and Gansu provinces and the ethnic and cultural advantages of the Ningxia Hui Autonomous Region and Qinghai Province, build Xi'an into a new focus of reform and opening-up in China's interior, speed up the development and opening-up of cities such as Lanzhou and Xining, and advance the building of the Ningxia Inland Opening-up Pilot Economic Zone with the goal of creating strategic channels, trade and logistics hubs and key bases for industrial and cultural exchanges opening to Central, South and West Asian countries. We should give full play to Inner Mongolia's proximity to Mongolia and Russia, improve the railway links connecting Heilongjiang Province with Russia and the regional railway network, strengthen cooperation between China's Heilongjiang, Jilin and Liaoning provinces and Russia's Far East region on sea-land multi-modal transport, and advance the construction of an Eurasian high-speed transport corridor linking Beijing and Moscow with the goal of building key windows opening to the north.

Southwestern region. We should give full play to the unique advantage of Guangxi Zhuang Autonomous Region as a neighbor of ASEAN countries, speed up the opening-up and development of the Beibu Gulf Economic Zone and the Pearl River-Xijiang Economic Zone, build an international corridor opening to the ASEAN region, create new strategic anchors for the opening-up and development of the southwest and mid-south regions of China, and form an important gateway connecting the Silk Road Economic Belt and the 21st-Century Maritime Silk Road.

We should make good use of the geographic advantage of Yunnan Province, advance the construction of an international transport corridor connecting China with neighboring countries, develop a new highlight of economic cooperation in the Greater Mekong Sub-region, and make the region a pivot of China's opening-up to South and Southeast Asia. We should promote the border trade and tourism and culture cooperation between Tibet Autonomous Region and neighboring countries such as Nepal.

Coastal regions, and Hong Kong, Macao and Taiwan. We should leverage the strengths of the Yangtze River Delta, Pearl River Delta, west coast of the Taiwan Straits, Bohai Rim, and other areas with economic zones boasting a high level of openness, robust economic strengths and strong catalytic role, speed up the development of the China (Shanghai) Pilot Free Trade Zone, and support Fujian Province in becoming a core area of the 21st-Century Maritime Silk Road. We should give full scope to the role of Qianhai (Shenzhen), Nansha (Guangzhou), Hengqin (Zhuhai) and Pingtan (Fujian) in opening-up and cooperation, deepen their cooperation with Hong Kong, Macao and Taiwan, and help to build the Guangdong-Hong Kong-Macao Big Bay Area. We should promote the development of the Zhejiang Marine Economy Development Demonstration Zone, Fujian Marine Economic Pilot Zone and Zhoushan Archipelago New Area, and further open Hainan Province as an international tourism island. We should strengthen the port construction of coastal cities such as Shanghai, Tianjin, Ningbo-Zhoushan, Guangzhou, Shenzhen, Zhanjiang, Shantou, Qingdao, Yantai, Dalian, Fuzhou, Xiamen, Quanzhou, Haikou and Sanya, and strengthen the functions of international hub airports such as Shanghai and Guangzhou. We should use opening-up to motivate these areas to carry out deeper reform, create new systems and mechanisms of open economy, step up scientific and technological innovation, develop new advantages for participating in and leading international cooperation and competition, and become the pace-setter and main force in the Belt and Road Initiative, particularly the building of the 21st-Century Maritime Silk Road. We should leverage the unique role of overseas Chinese and the Hong Kong and Macao Special Administrative Regions, and encourage them to participate in and contribute to the Belt and Road Initiative. We should also make proper arrangements for the Taiwan region to be part of this effort.

Inland regions. We should make use of the advantages of inland regions, including a vast landmass, rich human resources and a strong industrial foundation, focus on such key regions as the city clusters along the middle reaches of the Yangtze River, around Chengdu and Chongqing, in central Henan Province, around Hohhot, Baotou, Erdos and Yulin, and around Harbin and Changchun to propel regional interaction and cooperation and industrial concentration. We should build Chongqing into an important pivot for developing and opening up the western region, and make Chengdu, Zhengzhou, Wuhan, Changsha, Nanchang and Hefei leading areas of opening-up in the inland regions. We should accelerate cooperation between regions on the upper and middle reaches of the Yangtze River and their counterparts along Russia's Volga River. We should set up coordination mechanisms in terms of railway transport and port customs clearance for the

China-Europe corridor, cultivate the brand of "China-Europe freight trains," and construct a cross-border transport corridor connecting the eastern, central and western regions. We should support inland cities such as Zhengzhou and Xi'an in building airports and international land ports, strengthen customs clearance cooperation between inland ports and ports in the coastal and border regions, and launch pilot e-commerce services for cross-border trade. We should optimize the layout of special customs oversight areas, develop new models of processing trade, and deepen industrial cooperation with countries along the Belt and Road.

A.7 China in Action

For more than a year, the Chinese government has been actively promoting the building of the Belt and Road, enhancing communication and consultation and advancing practical cooperation with countries along the Belt and Road, and introduced a series of policies and measures for early outcomes.

High-level guidance and facilitation. President Xi Jinping and Premier Li Keqiang have visited over 20 countries, attended the Dialogue on Strengthening Connectivity Partnership and the sixth ministerial conference of the China-Arab States Cooperation Forum, and met with leaders of relevant countries to discuss bilateral relations and regional development issues. They have used these opportunities to explain the rich contents and positive implications of the Belt and Road Initiative, and their efforts have helped bring about a broad consensus on the Belt and Road Initiative.

Signing cooperation framework. China has signed MOUs of cooperation on the joint development of the Belt and Road with some countries, and on regional cooperation and border cooperation and mid- and long-term development plans for economic and trade cooperation with some neighboring countries. It has proposed outlines of regional cooperation plans with some adjacent countries.

Promoting project cooperation. China has enhanced communication and consultation with countries along the Belt and Road, and promoted a number of key cooperation projects in the fields of infrastructure connectivity, industrial investment, resource development, economic and trade cooperation, financial cooperation, cultural exchanges, ecological protection and maritime cooperation where the conditions are right.

Improving policies and measures. The Chinese government will integrate its domestic resources to provide stronger policy support for the Initiative. It will facilitate the establishment of the Asian Infrastructure Investment Bank. China has proposed the Silk Road Fund, and the investment function of the China-Eurasia Economic Cooperation Fund will be reinforced. We will encourage bank card clearing institutions to conduct cross-border clearing operations, and payment institutions to conduct cross-border payment business. We will actively promote investment and trade facilitation, and accelerate the reform of integrated regional customs clearance.

Boosting the role of cooperation platforms. A number of international summits, forums, seminars and expos on the theme of the Belt and Road Initiative have been held, which have played an important role in increasing mutual understanding, reaching consensus and deepening cooperation.

A.8 Embracing a Brighter Future Together

Though proposed by China, the Belt and Road Initiative is a common aspiration of all countries along their routes. China is ready to conduct equal-footed consultation with all countries along the Belt and Road to seize the opportunity provided by the Initiative, promote opening-up, communication and integration among countries in a larger scope, with higher standards and at deeper levels, while giving consideration to the interests and aspirations of all parties. The development of the Belt and Road is open and inclusive, and we welcome the active participation of all countries and international and regional organizations in this Initiative.

The development of the Belt and Road should mainly be conducted through policy communication and objectives coordination. It is a pluralistic and open process of cooperation which can be highly flexible, and does not seek conformity. China will join other countries along the Belt and Road to substantiate and improve the content and mode of the Belt and Road cooperation, work out relevant timetables and roadmaps, and align national development programs and regional cooperation plans.

China will work with countries along the Belt and Road to carry out joint research, forums and fairs, personnel training, exchanges and visits under the framework of existing bilateral, multilateral, regional and sub-regional cooperation mechanisms, so that they will gain a better understanding and recognition of the contents, objectives and tasks of the Belt and Road Initiative.

China will work with countries along the Belt and Road to steadily advance demonstration projects, jointly identify programs that accommodate bilateral and multilateral interests, and accelerate the launching of programs that are agreed upon by parties and ready for implementation, so as to ensure early harvest.

The Belt and Road cooperation features mutual respect and trust, mutual benefit and win-win cooperation, and mutual learning between civilizations. As long as all countries along the Belt and Road make concerted efforts to pursue our common goal, there will be bright prospects for the Silk Road Economic Belt and the 21st-Century Maritime Silk Road, and the people of countries along the Belt and Road can all benefit from this Initiative.

Appendix B
Promote Friendship Between Our People and Work Together to Build a Bright Future—Speech at Nazarbayev University

(Astana, 7 September 2013)
Speech by H.E. Xi Jinping, President of the People's Republic of China
Your Excellency President Nazarbayev,
President of Nazarbayev University,
Dear Faculty Members and Students,
Ladies and Gentlemen,
Dear Friends,

Good morning. I have come to Kazakhstan, China's great neighbor, on a state visit at the invitation of President Nazarbayev. It is my great pleasure today to come to Nazarbayev University and meet with all of you.

Let me begin by conveying the sincere greetings and best wishes of the Chinese people to the friendly people of Kazakhstan, the faculty members and students of this university and all the friends present today.

As a Kazakh proverb goes, "The history of a land is the history of its people." Since independence, Kazakhstan, under the leadership of President Nazarbayev, has seen sustained political stability, robust economic growth, significantly higher living standards and markedly growing international influence.

The city of Astana, where we are today, has grown into a beautiful, modern city in less than two decades. It is a success story written by the people of Kazakhstan on their magic land. Standing here, I could see both the hard work and ingenuity of the Kazakh people and the bright future ahead of them.

Ladies and Gentlemen,
Dear Friends,
Over 2100 years ago during China's Han Dynasty, a Chinese envoy Zhang Qian was sent to Central Asia twice with a mission of peace and friendship. His journeys opened the door to friendly contacts between China and Central Asian countries as well as the Silk Road linking east and west, Asia and Europe.

Shaanxi, my home province, is right at the starting point of the ancient Silk Road. Today, as I stand here and look back at that episode of history, I could almost hear the camel bells echoing in the mountains and see the wisp of smoke rising from the desert. It has brought me close to the place I am visiting.

W. Liu (ed.), *China's Belt and Road Initiatives*,
https://doi.org/10.1007/978-981-13-0101-8

Kazakhstan, sitting on the ancient Silk Road, has made an important contribution to the exchanges between the Eastern and Western civilizations and the interactions and cooperation between various nations and cultures. This land has borne witness to a steady stream of envoys, caravans, travelers, scholars and artisans travelling between the East and the West. The exchanges and mutual learning thus made possible have contributed to the progress of human civilization.

The ancient city of Almaty is also on the ancient Silk Road. In Almaty, there is a Xian Xinghai Boulevard, which got its name from a true story. At the outset of the Great Patriotic War in 1941, Xian, a renowned Chinese composer, arrived in Almaty after much travail. By then, he was worn down by poverty and illness and had no one to turn to. Fortunately, the Kazakh composer Bakhitzhan Baykadamov took care of Xian and provided him with the comfort of a home.

It was in Almaty that Xian composed his famous works Liberation of the Nation, Sacred War and Red All over the River. He also wrote a symphony Amangeldy based on the exploits of the Kazakh national hero. These pieces served as a rallying call to fight Fascists and proved immensely popular with the local people.

Throughout the millennia, the people of various countries along the ancient Silk Road have jointly written a chapter of friendship that has been passed on to this very day. The over 2000-year history of exchanges demonstrates that on the basis of solidarity, mutual trust, equality, inclusiveness, mutual learning and win-win cooperation, countries of different races, beliefs and cultural backgrounds are fully capable of sharing peace and development. This is the valuable inspiration we have drawn from the ancient Silk Road.

Ladies and Gentlemen,
Dear Friends,
Over the past 20 years and more, the relations between China and Eurasian countries have developed rapidly and the ancient Silk Road has gained fresh vitality. In a new way, it is taking the mutually beneficial cooperation between China and Eurasian countries to a new height.

A near neighbor is better than a distant relative. China and Central Asian countries are friendly neighbors connected by common mountains and rivers. China highly values its friendship and cooperation with these countries and takes them as a foreign policy priority.

At present, China's relations with Central Asian countries face a golden opportunity of development. We hope to work with these countries to strengthen trust, friendship and cooperation, and promote common development and prosperity so as to bring benefits to our people.

- We need to pass on our friendship from generation to generation and always be good neighbors living in harmony. China is committed to the path of peaceful development and the independent foreign policy of peace. We respect the development paths and domestic and foreign policies chosen independently by the people of every country. We will in no circumstances interfere in the internal affairs of Central Asian countries. We do not seek to dominate regional affairs or

establish any sphere of influence. We stand ready to enhance communication and coordination with Russia and all Central Asian countries to strive to build a region of harmony.
- We need to firmly support and trust each other and be sincere and good friends. To render each other firm support on major issues concerning core interests such as sovereignty, territorial integrity, security and stability is the essence and an important part of China's strategic partnership with Central Asian countries. We will deepen trust and cooperation with Central Asian countries bilaterally and within the Shanghai Cooperation Organization (SCO) to combat the "three forces" of terrorism, separatism and extremism as well as drug trafficking and transnational organized crime to create a favorable environment for the economic development and the well-being of the people in this region.
- We need to vigorously enhance practical cooperation and be good partners of win-win cooperation. Both China and Central Asian countries are at a crucial stage of development with unprecedented opportunities and challenges. We have all set medium- to long-term development goals based on our national conditions. Our strategic goals are the same, which is to ensure sustained and stable economic development, build a prosperous and strong nation and achieve national revitalization. Therefore, we need to enhance practical cooperation across the board, turn our good political relations, geographical proximity and economic complementarity into drivers of practical cooperation and sustained growth, and build a community of shared interests and mutual benefit.
- We need to expand regional cooperation with a more open mind and broader vision and achieve new glories together. The world is going through faster economic integration and regional cooperation is booming. The Eurasian region has quite a few regional cooperation organizations. The members and observers of the Eurasian Economic Community (EAEC) and the SCO are from Eurasia, South Asia and West Asia. By intensifying cooperation between the SCO and the EAEC, we will create even more space for development.

Ladies and Gentlemen,
Dear Friends,
To forge closer economic ties, deepen cooperation and expand development space in the Eurasian region, we should take an innovative approach and jointly build an "economic belt along the Silk Road". This will be a great undertaking benefitting the people of all countries along the route. To turn this into a reality, we may start with work in individual areas and link them up over time to cover the whole region.

First, we need to step up policy communication. Countries should have full discussions on development strategies and policy response, work out plans and measures for advancing regional cooperation through consultation in the spirit of seeking common ground while reserving differences, and give the policy and legal "green light" to regional economic integration.

Second, we need to improve road connectivity. The SCO is working on an agreement on transportation facilitation. If signed and implemented at an early date, it will open up a major transportation route connecting the Pacific and the Baltic

Sea. Building on that, we will actively discuss the best way to improve cross-border transportation infrastructure and work toward a transportation network connecting East Asia, West Asia and South Asia to facilitate economic development and travel in the region.

Third, we need to promote unimpeded trade. The proposed "economic belt along the Silk Road" is inhabited by close to 3 billion people and represents the biggest market in the world with unparalleled potential. The potential for trade and investment cooperation between the relevant countries is enormous. We should discuss a proper arrangement for trade and investment facilitation, remove trade barriers, reduce trade and investment cost, increase the speed and quality of regional economic flows and achieve win-win progress in the region.

Fourth, we need to enhance monetary circulation. China and Russia have already had sound cooperation on settling trade in local currencies, and have gained gratifying results and rich experience. There is no reason not to share this good practice with others in the region. If our region can realize local currency convertibility and settlement under current and capital accounts, it will significantly lower circulation cost, increase our ability to fend off financial risks and make our region more competitive economically in the world.

Fifth, we need to increase understanding between our people. Amity between the people holds the key to good relations between states. To have productive cooperation in the above-mentioned areas, we need the support of our people. We should encourage more friendly exchanges between our people to enhance mutual understanding and traditional friendship and build strong public support and a solid social foundation for regional cooperation.

Dear Faculty Members and Students,

Young people are the future of a nation. The great Kazakh poet and philosopher Abay Qunanbayev once noted, "The world is like an ocean and our time is like strong wind. Waves in the front are the cider brother while those behind are the younger brother. Driven by wind, the waves from behind constantly press on those in the front. This has been the case since ancient times." Seeing so many young glowing faces in the audience, I could not but recall with fond memories my own life in university. It was such an unforgettable time of youth!

The Kazakh people often say, "Knowledge brings light to the world; without it, confusion will reign." Indeed, knowledge is power. Young people are in the very prime of life, the best time to acquire knowledge, build character and develop ability. I am sure that the young people to graduate from this university will become future pillars in Kazakhstan's national revitalization.

To facilitate youth exchanges within the SCO framework, China will, in the coming ten years, provide 30,000 government scholarships to SCO member states and invite 10,000 teachers and students from Confucius Institutes in these countries to China for study tours. I hope you will seize these opportunities to come to China for study and exchange.

I also wish to invite 200 faculty members and students from Nazarbayev University to go to China next year and attend summer camp activities.

Ladies and Gentlemen,
Dear Friends,
Young people are the mainstay of friendship between our people. The common interest and ideal they hold for life all bring them together in true friendship. Here let me share with you two touching stories involving Chinese and Kazakh people.

Back in the early 1950s, a Chinese young man working in Xinjiang met a pretty girl called Valentina at a local hospital. They fell in love, got married and had a son. However, due to the political climate and for other reasons, Valentina returned to her home country when their son was only six years old. When the boy grew up, he tried all means to find his mother but to no avail. In 2009, at the age of 61, he finally found out that his mother was living in Almaty. He visited his 80-year-old mother there and took her to China for sightseeing. This happy family reunion, though overdue for half a century, is a strong testament to the friendship between our people.

The other story is about a young Kazakh student whose name is Ruslan. You may know that the RH-negative blood type is called "panda blood" in China because it is as scarce as a panda. It is very hard for RH-negative patients to find compatible blood source. Ruslan, who studied at Hainan University in China, happens to be RH-negative. Since 2009, he has been donating blood twice a year, helping to save Chinese patients of his same blood type. When praised by his Chinese friends, he simply said, "I feel this is what I should do. It is my duty to help others."

The stories 1 have just told are two episodes from the epic of friendly exchanges between our people, yet they fully demonstrate how close the Chinese and Kazakh people are to each other.

I believe that young people from China and Kazakhstan, including those of you in the audience, are the future envoys of friendship between our two countries. You will contribute your youth and energy to the growth of China-Kazakhstan comprehensive strategic partnership.

Ladies and Gentlemen,
Dear Friends,
China and Kazakhstan are friendly neighbors as close as lips and teeth. Our 1700-km long common border, two millennia of interactions and extensive common interests not only bind us closely together, but also promise a broad prospect for bilateral ties and mutually beneficial cooperation. Let us join hands to carry on our traditional friendship and build a bright future together.

Thank you.

Appendix C
Join Hands to Build the China-ASEAN Community of Common Destiny—Speech at the Indonesian Parliament

(Jakarta, 3 October 2013)
Speech by H. E. Xi Jinping, President of the People's Republic of China

Honorable Speaker Marzuki Alie and deputy speakers of the House of Representatives,
Honorable Chairman Sidarto Danusubroto and vice-chairmen of the People's Consultative Assembly of the Republic of Indonesia,
Honorable Speaker Irman Gusman and deputy speakers of the Regional Representative Council, speakers, ministers,

Ladies and gentlemen,
Dear Friends,
Apa Kabar! Good morning. It gives me great pleasure to come to the People's Representative Council of Indonesia and meet all the friends here. I am visiting Indonesia, known as "country of thousands islands", at the invitation of President Susilo Bambang Yudhoyono. This is the first leg in my current visit to Southeast Asia. It is a journey for carrying forward traditional friendship and also for planning future cooperation.

I visited Indonesia 20 years ago, during which I personally experienced the development, colorful scenery and diversified culture of Indonesia. How time flies. But what I saw 20 years ago is still vivid in my mind, as if it only happened yesterday. As I set foot on this beautiful land again this time, I am even more impressed by the robust vitality of our bilateral relations and the profound friendship between the two peoples. In recent years, under the leadership of President Yudhoyono, the Indonesian people have pulled together, made tremendous efforts and achieved success in economic development, social stability and growing national strength. I sincerely wish the Indonesian people further success in forging a better future through hard work and creativity.

W. Liu (ed.), *China's Belt and Road Initiatives*,
https://doi.org/10.1007/978-981-13-0101-8

Ladies and Gentlemen,
Dear Friends,
China and Indonesia face each other across the sea. The friendly ties between us have a long history. Together, our peoples have composed one piece after another of beautiful music about their exchanges and interactions over the centuries. Just as the Indonesian folk song Bengawan Solo, a household musical piece in China, goes "Your water springs forth from Solo, caged by a thousand mountains. Water flows to reach far distances, eventually to the sea." Like the beautiful river Solo, China-Indonesia relations have traversed an extraordinary journey, past mountains and eventually to the sea. As early as the Han Dynasty in China about 2000 years ago, the people of the two countries opened the door to each other despite the sea between them. In the early 15th century, Zheng He, the famous Chinese navigator of the Ming Dynasty, made seven voyages to the Western Seas. He stopped over the Indonesian archipelago in each of his voyages and toured Java, Sumatra and Kalimantan. His visits left nice stories of friendly exchanges between the Chinese and Indonesian peoples, many of which are still widely told today.

Over the centuries, the vast oceans have served as the bond of friendship connecting the two peoples, not a barrier between them. Vessels full of goods and passengers travelled across the sea, exchanging products and fostering friendship. A Dream of Red Mansions, a Chinese classic novel, gives vivid accounts of rare treasures from Java. The National Museum of Indonesia, on the other hand, displays a large number of ancient Chinese porcelains. All these bear witness to the friendly exchanges between the two peoples. And they are convincing interpretation of the Chinese saying that "A bosom friend afar brings a distant land near." Our two peoples sympathized with and supported each other in their respective struggle for national independence and liberation in the last century. Indonesia was among the first countries to establish diplomatic ties with the People's Republic of China after its founding in 1949. In 1955, China and Indonesia, together with other Asian and African countries, jointly initiated the Bandung spirit at the Bandung Conference. With the principles of peaceful coexistence and seeking common ground while shelving differences at its core, the Bandung spirit remains an important norm governing state-to-state relations, and has made indelible contribution to the building of new international relations. Our two countries resumed diplomatic ties in 1990 and established a strategic partnership in 2005, which marked the start of a new phase in the growth of our bilateral relations.

Ladies and Gentlemen,
Dear Friends,
During my current visit, President Yudhoyono and I have jointly announced our decision to upgrade our bilateral relations to a comprehensive strategic partnership, with a view to building on past achievements and bringing about all-round and in-depth growth of our relationship. The growing mutual trust between the two countries has put our bilateral relations on a more solid political basis. Our practical cooperation has expanded from such traditional areas as economy, trade, finance,

infrastructure, energy, resources and manufacturing to include new fields such as space and maritime affairs. Our wide-ranging cooperation, from space to ocean, has delivered tangible benefits to the two peoples. The Surabaya–Madura Bridge, a China-Indonesia joint project, is the longest cross-sea bridge in Southeast Asia. The cooperation project of Jatigede Dam is about to finish. Once completed, it will have an irrigation area of 90,000 ha, which will greatly facilitate the work and life of the local people. We have enhanced our cooperation on major regional and international issues. Our relations have increasingly gained regional and international influences. All this is of positive significance to the building of a more just and equitable international political and economic order. All these are important landmarks of China-Indonesia friendship in the new era.

People in Indonesia often say, "It's easy to make money but difficult to make friends." The sincere friendship between the Chinese and Indonesian people is indeed our cherished, priceless treasure. On 26 December 2004, a sudden massive earthquake of magnitude 9 hit the once tranquil Indian Ocean and triggered a large tsunami. The whole world was shocked to see Aceh of Indonesia suffering huge losses of life and property. After the tsunami, China immediately activated the emergency response mechanism and announced on the day of the earthquake that it would provide assistance to Indonesia and other disaster-affected countries. It became the largest overseas rescue and relief operation that China had ever launched after the founding of the People's Republic. Trucks carrying relief supplied were given green lights all the way and rushed from factories to airport, and airplanes, loaded with the love and sympathy of the Chinese people toward the Indonesian people, flew to Aceh and other disaster-hit areas. The Chinese international rescue team was the first foreign team to arrive at Aceh. They provided medical treatment to over 10,000 victims in a short span of 13 days. Many local people learned to speak Chinese and hailed members of the Chinese rescue team in Chinese, "zhongguo, beijing, wo ai ni" (China, Beijing, I love you).

The Chinese people, too, offered their sympathy and assistance in a variety of ways to the Indonesia people hit by the disaster. An old man in Hangzhou, who was not rich himself as his wife was hospitalized for a long time and he himself had just gone through a surgery, donated 1000 yuan of his hard earned money to Indonesian kids in disaster-affected areas for their continued schooling. The donation, though not a bit sum, speaks to the profound goodwill of the Chinese people to the Indonesian people. Likewise, the Indonesian people have extended a helping hand when the Chinese people experienced major natural disasters. On 12 May 2008, a massive earthquake hit Wenchuan, China. People in the affected areas were in dire need of assistance and relief. Indonesia responded immediately and sent its medical team to quake-hit areas. As soon as it arrived, the Indonesian medical team worked around the clock in spite of the many aftershocks. They treated 260 earthquake victims and provided free medical treatment to 844 local residents and 120 students. Before returning to Indonesia, they donated all their valuable belongings to the quake-affected areas. Back in Indonesia, people made donations both in cash and kind to the quake-stricken Wenchuan. Some Indonesians personally visited the Chinese Embassy in Indonesia to express their prayers and blessings to the people in Wenchuan. What they did has deeply moved the Chinese people. There are countless touching stories like these about friendly ties between our peoples. They are all examples of what is captured by one common proverb in both the Chinese and Indonesian languages, namely "going through thick and thin together".

Ladies and Gentlemen,
Dear Friends,
China and ASEAN countries are linked by common mountains and rivers and share a historical bond. This year marks the tenth anniversary of the China-ASEAN strategic partnership. Our relationship now stands at a new historical starting point. China places great importance on Indonesia's status and influence in ASEAN. China wishes to work with Indonesia and other ASEAN countries to ensure that China and ASEAN are good neighbors, good friends and good partners who would share prosperity and security and stick together through thick and thin. By making joint efforts, we will build a more closely-knit China-ASEAN community of common destiny so as to bring more benefits to both China and ASEAN and to the people in the region. To that end, we should focus our efforts in the following areas:

First, build trust and develop good-neighborliness. Trust in the very foundation of both interpersonal and state-to-state relations. China is committed to a relationship of sincerity and friendship with ASEAN countries and to enhanced mutual political and strategic trust. There is no one-size-fits-all development model in the world or an unchanging development path. Both the Chinese people and people in ASEAN countries have embraced change and innovation with an open mind, and explored and found, in a pioneering and enterprising spirit, development paths in light of their specific national conditions that conform to the trend of times. All these efforts have opened up a broad prospect for their economic and social development. We should respect each other's right to independently choose social system and development path as well as each other's efforts to explore and pursue economic and social development, and improve people's lives. We should have full confidence in each other's strategic direction, support each other on issues of major concern, and never deviate from the general direction of China- ASEAN strategic cooperation. China is ready to discuss with ASEAN countries the prospect of concluding of a treaty of good-neighborliness, friendship and cooperation in a joint effort to build good-neighborly relations. China will continue to support ASEAN in growing its strength, building ASEAN community and playing a central role in regional cooperation.

Second, work for win-win cooperation. As a well-known Chinese saying goes, "The interests to be considered should be the interests of all," China is ready to open itself wider to ASEAN countries on the basis of equality and mutual benefit to enable ASEAN countries to benefit more from China's development. China is prepared to upgrade the China-ASEAN Free Trade Area and strive to expand two-way trade to one trillion US dollars by 2020. China is committed to greater connectivity with ASEAN countries. China will propose the establishment of an Asian infrastructure investment bank that would give priority to ASEAN countries' needs. Southeast Asia has since ancient times been an important hub along the ancient Maritime Silk Road. China will strengthen maritime cooperation with ASEAN countries to make good use of the China-ASEAN Maritime Cooperation Fund set up by the Chinese government and vigorously develop maritime partnership in a joint effort to build the Maritime Silk Road of the 21st century. China is ready to expand its practical cooperation with ASEAN countries across the board, supplying each other's needs and complementing each other's strengths, with a view to jointly seizing opportunities and meeting challenges for the benefit of common development and prosperity.

Third, stand together and assist each other. China and ASEAN countries are as close as lips and teeth. We share the responsibility for regional peace and stability. Historically, the people of China and ASEAN countries had stood together through thick and thin in the fight to take our destiny back into our own hands. In recent years, our peoples have stood side by side and forged strong synergy in responding to various crises from Asian financial crisis to the international financial crisis, and from the Indian Ocean tsunami to China's Wenchuan earthquake. We should cast away the Cold War mentality, champion the new thinking of comprehensive security, common security and cooperative security and jointly uphold regional peace and stability. We should deepen cooperation in disaster prevention and relief, cyber security, combating cross-border crimes and joint law enforcement to create a more peaceful, tranquil and amicable home for the people in the region. China is ready to work with ASEAN countries to improve China-ASEAN defense ministers' meeting mechanism and hold regular dialogues on regional security issues. With regard to differences and disputes between China and some Southeast Asian nations on territorial sovereignty and maritime rights and interests, peaceful solutions should be sought, and differences and disputes should be properly handled through equal-footed dialogue and friendly consultation in the overall interests of bilateral ties and regional stability.

Fourth, enhance mutual understanding and friendship. Just as a Chinese saying goes, "A big tree grows from a small seedling; and a nine-storeyed tower is built out of soil." To ensure that the tree of China-ASEAN friendship remains evergreen, the soil of social support for our relations should be compacted. Last year saw 15 million people traveling between China and ASEAN countries with over 1000 flights between the two sides each week. Increased interactions have nurtured deeper bonds between us and enabled our people to feel ever closer to each other. We should encourage more friendly exchanges between youth, think tanks, parliaments, NGOs and civil organizations of the two sides, which in turn will generate more intellectual support to the growth of China-ASEAN relations and help increase the mutual understanding and friendship between our peoples. China is ready to send more volunteers to ASEAN countries to support their cultural, educational, health and medical development. China has proposed to set 2014 as the year of China-ASEAN cultural exchanges. In the coming three to five years, china will offer ASEAN countries 15,000 government scholarships.

Fifth, stick to openness and inclusiveness. The sea is big because it admits all rivers. In the long course of human history, the people of China and ASEAN countries have created splendid and great civilizations renowned around the world.

Ours is a diversified region. Various civilizations have assimilated and interacted with one another under the influence of different cultures, which affords and important cultural foundation for the people China ASEAN countries to learn from and complement one another. We should draw upon the experience of other regions in development and welcome a constructive role by non-regional countries in the development and stability of the region. At the same time, these, non-regional countries should respect the diversity of the region and do more to facilitate regional development and stability. The China-ASEAN community of shared destiny is closely linked with the ASEAN community and the East Asia community. The two sides need to bring out their respective strengths to realize diversity, harmony, inclusiveness and common progress for the benefit of the people of the region and beyond. A more closely knit China-ASEAN common destiny conforms to the trend of the times for seeking peace, development, cooperation and mutual benefit, and meets the common interests of the people of Asia and the world, hence enjoying a broad space and huge potential of growth.

Ladies and Gentlemen: Dear Friends, Since the founding of the People's Republic of China over 60 years ago, in particular the launch of reform and opening-up over 30 years ago, China has blazed a successful path of development and made tremendous progress. China has laid out a strategic plan and set clear goals for its future development, namely, to double its 2010 GDP and per capita income of its urban and rural residents and finish the building of a moderately prosperous society in all respects by 2020; and turn China into a modern socialist country that is prosperous, strong, democratic, culturally advanced and harmonious and realize the great renewal of the Chinese nation by the middle of this century. This is a dream that has long been cherished by the Chinese nation and people. It is also a prerequisite for China to make even greater contribution to mankind. As an ancient Chinese maxim goes, "only with high ambition and hard work can one make great achievements." We have the confidence, conditions and capabilities to attain our goals. Having said that, we are soberly aware that China remains the largest developing country in the world. We still face many difficulties and challenges on the way ahead. For all the Chinese people to enjoy a happy life, we have to make unrelenting efforts for a long time to come. We will firmly stay on the course of reform and opening-up, adhere to the path of socialism with Chinese characteristics, focus on managing our own affairs well, press ahead with modernization and improve people's wellbeing. China cannot achieve development in isolation from the world, and the world also needs China for development. China is fully committed to the path of peaceful development, the independent foreign policy of peace and the opening-up strategy for win-win results. A stronger China will add to the force for world peace and the positive energy for friendship, and will present development opportunities to Asia and the world, rather than posing a threat. China will continue to share opportunities for economic and social development with AESAN, Asia and the world.

Ladies and Gentlemen,
Dear Friends,
The Chinese people are working hard to realize the Chinese dream for the great renewal of the Chinese nation. And here, the Indonesian people are vigorously implementing their overall economic development plan for the rise of Indonesian nation. To realize our respective dreams, it is all the more necessary for both sides to understand, support and cooperate with each other, and for more people of vision in both countries to get involved and make effective efforts to this end.

In this connection, I recall the song Hening (Silence) composed by President Yudhoyono. In October 2006, he came to Guangxi, China for the commemorative summit marking the 15th anniversary of China-ASEAN dialogue relations. On the Lijiang River, President Yudhoyono was overwhelmed by inspiration and wrote down these beautiful lyrics. "The beautiful days I spend with my friends have kept recurring in my life." The mountains and rivers in China deeply touched President Yudhoyono and reminded him of his childhood and hometown. This shows just how strong the bond and affinity are between our two peoples.

Amity between people holds the key to good relations between nations. It is through the efforts of these envoys of friendship who have built bridges of friendship and opened windows to sincere understanding that the friendship between our two peoples has been everlasting and has grown stronger and more robust as time goes by. Young people are full of dynamism and dreams. When the young prosper and are strong, the country will prosper and be strong. Young people represent the future and hope for exchanges between the two countries. President Yudhoyono and I have agreed that the two countries will expand and deepen cultural and people-to-people exchanges. In the next five years, each side will send 100 young people to visit the other every year, and China will provide 1000 scholarships to Indonesia.

I am convinced that with more and more young people involved in the building of friendship between the two countries, China-Indonesia friendship will flourish and pass on from generation to generation.

Ladies and Gentlemen,
Dear Friends,
China and Indonesia have a combined population of 1.6 billion. As long as our two peoples work hand in hand with one heart and one mind, we will pool the huge strength of one fourth of mankind and be able to create new miracles in the history of human development. Let our two peoples join hands, write a new chapter in growing our bilateral relations, work for a bright future for the China-ASEAN community of shared destiny, and make greater contribution to the lofty cause of world peace and development.

Terima Kasih! (Thank you!)

FSC
www.fsc.org
MIX
Papier | Fördert
gute Waldnutzung
FSC® C083411

Kolibri 360 GmbH
Ferdinand-Jühlke-Straße 7
99095 Erfurt, Deutschland
produktsicherheit@kolibri360.de